Producer organisations and market chains

partageons les connaissances au profit des communautés rurales

sharing knowledge, improving rural livelihoods

About CTA

The Technical Centre for Agricultural and Rural Cooperation (CTA) was established in 1983 under the Lomé Convention between the ACP (African, Caribbean and Pacific) Group of States and the European Union Member States. Since 2000, it has operated within the framework of the ACP-EU Cotonou Agreement. CTA's tasks are to develop and provide products and services that improve access to information for agricultural and rural development, and to strengthen the capacity of ACP countries to acquire, process, produce and disseminate information in this area.

CTA is financed by the European Union.

CTA
Postbus 380
6700 AJ Wageningen
The Netherlands
Website: www.cta.int

Producer organisations
and market chains

Facilitating trajectories of change in developing countries

edited by:
Giel Ton
Jos Bijman
Joost Oorthuizen

Wageningen Academic
P u b l i s h e r s

This publication benefited from a financial contribution of the Netherlands Ministry of Agriculture, Nature and Food Quality (LNV).

ISBN 978-90-8686-048-7

First published, 2007
Reprint, 2010

Omslagontwerp:
Erik Vos - Het Lab, Arnhem
Illustratie: Anka Kresse, Arnhem

©Wageningen Academic Publishers
The Netherlands, 2010

Table of contents

Producer organisations and market chains

Introduction

Giel Ton, Jos Bijman and Joost Oorthuizen

> *'Though it is often considered as a solution, organisations are always a problem and rural producers' organisations are not an exception to the rule as testified by the many tensions that characterize their life and the "weariness" often felt by their leaders'* (Mercoiret 2006).

1. Why this book?

The role of producer organisations (POs) in market chains has received increasing attention in recent years, both from governments and donors. For instance, in 2002, the World Bank emphasized the importance of agricultural cooperatives and formal business associations for market support functions (policy development and advocacy) as well as market complementing and market substituting functions (e.g. setting up factor markets, developing alternative marketing channels, engaging in supply chains) (Bosc *et al.*, 2001; World Bank, 2002). In the upcoming 2008 World Development Report, the Bank stresses the importance of governments and donors in terms of enhancing the effectiveness of producer organisation participation in consultative policy processes. However, though the policy openings for support seem promising, smallholder market access through farmer-led economic organisations is not easy. In order to lower transactions costs, markets demand that smallholder farmers operate in an organised manner. However, such organising processes are not easy to achieve and development agencies often end up pushing smallholder organisations to 'overextend themselves' by failing to fully recognize the constraints (Hellin *et al.*, 2006).

This book is the result of a Dutch partnership between policy makers, researchers and practitioners designed to confront ideas with realities. Organised in a platform called Agri-ProFocus, members aim to provide more and better support to producer organisations in the South. Members of Agri-ProFocus work together in chain development programmes to provide more integral and comprehensive support to producer organisations. Through so-called expert meetings, staff from donor organisations and knowledge centres, government officials and business representatives share their experiences and lessons learned.

We think this book is timely because it not only underlines the need for support to producer organisations in developing countries, but also provides a panoramic view of the difficulties related to support activities. The experiences presented in this book are not recipes for instant success, but instead, highlight a trajectory for change that is often more fragile and slower moving than policy makers realise. This book contains experiences presented in expert meetings held in 2005 and 2006. The first meeting, entitled: 'Working with producer organisations: a welcome challenge or a nightmare?' brought together credit organisations, development NGOs, policy makers, researchers, training institutes and development practitioners. It soon became clear that participants were eager to share experiences because they were all struggling to find appropriate support structures for farmers' organisations. The subsequent expert meetings focussed on more specific issues, like legal frameworks, training needs and credit mechanisms (see www. agri-profocus.nl for details of these meetings).

2. Thematic content of the book

We decided to organise contributions in this volume into three thematic parts. This division is somewhat arbitrary; many contributions have content that could well be placed under different thematic headings. To illustrate the thematic divisions and get an idea of the difficulties producer organisations face in creating viable value chains, we will begin with an account from an economic producer organisation.

Imagine yourself as Juan Zamora in the following story:
Juan Zamora is in his office at the dairy processing plant of Negowa. The plant was constructed with state support in the 1970s, privatised as a shareholder firm in the late 1990s, and was bought by the milk producers some years ago with technical support from a development NGO. But its tough going. Sales are steady, but every year losses are incurred. Last year the bank took the plant as collateral. Payment on schedule is an ongoing problem.

Last year Juan was elected to be president of the board, and he now prepares for one of the toughest board meetings yet. To make the plant break even, he will have to fire half of the staff, including the manager, and lower prices paid to shareholder-members. This is highly disturbing. Some of the people he will have to fire are fellow villagers and his relatives. And if that's not enough, lower farm gate prices will directly affect his own income as well as that of his fellow producers. These are hard times.

Juan takes a deep breath and makes up his mind. The consultants painted a clear picture. But are their economic projections of milk sales really credible? They also made another point: that firing employees is different than firing agricultural labourers. Workers claim one month of compensation for every year worked! So instead of paying the monthly payroll, they must be paid almost one year's additional income ... and that's for NOT working!!! It is overwhelming: conflicting claims and conflicting loyalties. Juan does not know who to trust. The consultant, and all of those experts who are in no way financially affected by the decision, have taken pretty rigid positions ..... yet they all have secure jobs and solid saving accounts. His future son-in-law is amongst those he will have to fire. His daughter is distraught about it and threatens to move out. 'Why do I do this?' he think to himself, 'Why not wait until the next elections and leave this decision for my successor. It seems I can only lose.'

Juan reflects. A ray of hope begins to surface: 'We have been offered a loan from an NGO that will help us cover legal compensation for everyone being laid off. Perhaps we can convince them to give us a two year grace period?' He recalls a conversation with a fellow farmer leader he met at a meeting who explained an eye-opening management concept to him: 'Better half of the workers get laid off this year, than all of them in five year.' Better a short-term sacrifice on farm gate prices than the complete elimination of marketing possibilities in five years' time. 'I am not cut out to be a manager', he thinks, 'but I will have to do what I am elected for...' Juan makes a decision and then steps into the meeting room.

The story highlights a critical moment of decision making and entrepreneurship that determines sustainable access to markets for a specific group of smallholder farmers. It also illustrates the complexity of management decisions in economic producer organisations. Some may argue that these types of decisions are better taken by a professional manager, rather than by a self-made farmer leader. Western European cooperatives have evolved by separating responsibilities between a board of directors and professional management. However, in the context of smallholder production in developing countries, the costs of professional management are often too high relative to the 'margin' gained by the PO's economic activities. Hence, self-management is often the most feasible option when external support is absent. The story illustrates the issue of *organisational strengthening of producer organisations* and specific challenges associated with this support with view towards economic sustainability after external support ends.

Juan's decisions are not exclusively related to producer interests, but obey the logic of a small enterprise whose product must compete in the market with processed products. One of the key features of economic producer organisations is their drive to add value to farm gate products. This is done through different economic activities like processing, packaging, branding and by creating economies of scale in the initial stages of the value chain. Like other companies they are increasingly linked to other chain operators in *value chains*, where consumer demands and the quality requirements of retailers must be met by producers, and where marketing and innovation are key to economic success.

The story also indicates the diversity of stakeholders influencing Juan's decisions. He not only expresses the interests of fellow smallholder producers, but must also represent other stakeholders with interests vested in the plant. He has to interact and be a trustworthy partner for banks, NGOs and consultants who despite sharing a common interest (viability of the plant), do not necessarily have the same stake as farmers who supply the plant and whose direct earnings come from the farm gate dairy price, which represents the sourcing costs for the plant's raw material. This enabling *institutional environment* changes over time, and producer organisations will need to find adequate responses to these changes. They will react in a way that reflects their economic, cultural and political experiences - all of which will feed their assessment of the changes in the institutional environment, and influence decision making in terms of coping strategies.

The case studies included in this book have been organised around three broad thematic issues distilled from Juan's story: organisational support, value chain development and the institutional environment.

2.1 Section A. Organisational support for producer organisations

Providing organisational support for POs, while perhaps a welcome challenge, still requires skilled craftsmanship and a certain degree of good luck. Efforts made externally to support internal organisational change are necessary but can be frustrated by decision making procedures, political power struggles, or conflicting interests within the producer organisation. This section of the book presents several experiences with external support for the organisational strengthening of producer organisations. Gerrit Holtland describes the Naukat Seed Potato Growers Cooperative in the south of the Kyrgyz Republic. He describes how a balance was found between insiders (members, directors) and outsiders (NGOs and donors), as well as between organisational development and business development. Lithzy Flores, Dick Commandeur, Giel Ton and Gerda Zijm present a specific approach towards coaching POs in Bolivia, according to the different stages in their development. This methodology emerged from SNV's longstanding experience supporting smallholder organisations in the Andean Region of Bolivia. The authors describe the process and criteria that proved useful in the organisations' self-assessment, emphasizing market orientation, transparency and the generation of income to reduce donor dependency.

In her contribution, Olga van der Valk argues that culture is a decisive element in the organisational processes of producer organisations. A case study of coffee growing organisations in Mexico serves as an example of the dynamics influencing the relationship between producer organisations and external advisors, as part of an organisations' strategy to optimally interact with the external environment.

The last contribution of this section is from Kees Blokland and Christian Gouët who present a specific approach towards organisational support, that of peer-to-peer support between farmers' organisations. They describe several cases to illustrate the advantages of peer-to-peer exchange in terms of generating tangible benefits and services to members.

2.2 Section B. Value chain development with producer organisations

POs play an increasingly important role in national and international agrifood supply chains as intermediaries between individual farming households and chain actors such as buyers, processors and governments. They may have several functions, for example, collecting, processing and marketing agricultural products, implementing quality assurance programs, and providing members with advice and training. By exploiting economies of scale as well as by reducing transaction costs, POs can improve the efficiency and efficacy of agrifood supply chains. The social capital embedded in the PO addresses the traditional weakness of scattered smallholder production. Smallholder farmers need to be organised or they lose out to larger commercial farmers

Producer organisations and market chains

around the world, especially when it comes to accessing high value markets (supermarkets, fair trade, export, etc.).

We have clustered together several contributions that focus on these new challenges in the value chain. Jos Bijman discusses the challenges facing cooperatives in developed countries when partnering in (international) supply chains. After describing both traditional and modern functions of cooperatives, the author discusses a number of challenges that are particularly related to the (new) role of cooperatives in supply chains, focussing on financing, corporate governance, member commitment and member heterogeneity. Lucian Peppelenbos and Hugo Verkuijl present a framework to help farmers reflect on their position in value chains according to their level of involvement in chain activities, and their participation in chain management.

'Fair Trade' is a specific form of value chain where producer organisations have a special niche as producers and as partners in chain governance. The fair trade system generates different benefits for smallholders in developing countries: access to high value markets with a specific 'brand', a price premium that allows the farmers' organisation to invest in social projects in their communities, and a reduction in the price risk inherent in most commodity markets by offering a minimum price for specific commodities. The price premium for fair trade developed initially within the coffee sector. Dave Boselie analyses the recent introduction of fair trade standards into the fresh fruit sector, which began in 1996. It has proven to be a successful initiative to provide smallholder producers with an opportunity to access rapidly expanding supermarket retail segments in many European countries. However, the recent arrival of multinational companies in the fair trade arena threatens to push smallholders out of the international retail market once again. The author argues that co-ownership of the trading firm and the direct influence that these producers have on value chain development should guarantee the benefits of fair trade for small farmers. Bo van Elzakker assesses the benefits of fair trade and organic production in eight cases of export-oriented organic coffee and fruit production in East Africa. He details the costs and benefits of the process to source the product and export it to the European market. Donor commitment to subsidize these costs in a take-off period proves a necessary precondition for success in most of the projects reviewed.

The on-site monitoring of compliance to 'new' quality requirements for production that are inherent in fair trade and organic certification systems, is expensive. For smallholder farmers these costs are often prohibitive. Group certification via Internal Control Systems (ICS) is a potentially cost-effective way for smallholder POs to access market chains. Rhiannon Pyburn provides an in-depth analysis of several key challenges for the development of robust smallholder group certification systems. She argues that learning and developmental dimensions of an ICS are inextricably intertwined with credibility and inspection, and merit more attention. Issues related to documentation and defining the parameters of a given ICS are also addressed. Myrtille Danse presents a Costa Rican case where external support to cooperatives to meet specific quality demands has triggered a process of learning and continuous improvement of production.

Bertus Wennink and Ted Schrader describe the rebuilding of the potato supply chain and the roles of different stakeholders in Rwanda. The authors particularly focus on initiatives taken by producer organisations to develop the supply chain, and the challenges they experience in linking grassroots members with research institutes and other chain stakeholders. In the final chapter of this section, Joost Nelen paints a picture of cotton farmers' organisations in West Africa. He presents three examples to demonstrate how farmers' organisations, of their own initiative, have gradually come to grips with the rather complex issues plaguing the cotton sector.

2.3 Section C. Changes in the institutional environment for producer organisations

It is clear that the potential of POs to provide economic benefits for members is partly determined by the 'room for manoeuvre' provided by institutional environments. The institutional environment - the amalgam of rules and institutions in a society that define the 'rules of the game' for producer organisations - facilitates some types of POs and constrains others. Many POs have found ways to organise themselves to confront these constraints and make the best of it.

In this section, Kees Blokland and Christian Gouët provide the theoretical background of Agriterra's approach to support farmers' organisations and highlight the need for further research to pinpoint the policies, rules and institutions (the institutional environment) that can free this potential for economic development and democratization and that may help to replicate or upscale successful experiences. Jos Bijman, Rik Delnoye and Giel Ton review recent changes in the institutional environment in China that have given rise to new types of

POs. A new Cooperative Law will play a crucial role in creating more enabling institutional conditions for up-scaling recent pilot experiences in different Chinese provinces. The authors point out ambivalence in the law and warn against top-down government support that could strangle bottom-up farmer-led initiatives. In the subsequent chapter, Giel Ton analyzes the increasing involvement of farmer organisations in agricultural R&D support. He points to the growing importance of competitive funds for agricultural R&D, and the growing role of farmers' federations as mediators between grassroots organisations and private service providers in governance of the contracting process and contract conditions with researchers and extension workers. He illustrates these challenges for farmers' federations with examples from two R&D programs in Bolivia. Sietze Vellema concentrates on the role of producers organisations in the Philippines, particularly in contract farming schemes. He points to the historical and ideological background of farmers' organisations that explain that, although contract farming is highly debated in the Philippine peasant movement, these organisations have to date played a minor role in assisting the involved farmers in contract negotiations and other governance issues in the contract farming value chain. He points out that a win-win situation wherein producer organisations influence and manage the terms of the contract, is not an automatic outcome of 'contracted' market access.

3. Ways forward

Contributions to this book illustrate that local, regional and international market chains are developing rapidly around the globe, putting high demands on producers to access these chains and meet related product or process requirements. The danger of smallholder farmers being marginalized in the process is very real. POs may counter marginalization by organising farmers for improved production, lower transaction costs and giving them a stronger voice towards the outside. But they need support in order for this to happen. Government support in creating an enabling environment for economic activities, and support for increasing capacities to adapt activities to changing conditions.

Agri-ProFocus believes that strong POs are the key to rural economic development and liveable rural areas. Cooperation between member organisations focusing on agrarian producers, contributes to this cause. Agri-ProFocus strives for effective cooperation between Dutch member organisations. Through consultancy, training, research, marketing, finance, and credit provision, the partnership endeavors to offer powerful, coherent, and problem-orientated support to rural producer organisations.

Wageningen UR is working to establish a thematic research area around producer organisations and value chains in developing countries. Together with the Royal Tropical Institute and the Radboud University Nijmegen, the two other research institutes comprising Agri-ProFocus, Wageningen UR will contribute to a knowledge platform where conceptual frameworks and the writing skills of researchers meet the practical knowledge and real-life accounts of practitioners and farmers. As editors we hope that this volume feeds this quest for knowledge, experiences and lessons learned of all those interested in supporting and facilitating trajectories of change led by producer organisations in developing countries.

References

Bosc, P., D. Eychenne, K. Hussein, B. Losch, M.-R. Mercoiret, P. Rondot and S. Mackintosh-Walker, 2001. Reaching the Rural Poor: The Role of Rural Producers Organisations (RPOs) in the World Bank Rural Development Strategy - background study, World Bank.

Hellin, J., M. Lundy and M. Meijer, 2006. Farmer Organisation, Collective Action and Market Access in Meso-America. Research Workshop on Collective Action and Market Access for Smallholders, Cali, Colombia.

Mercoiret, M.-R., 2006. Rural Producer Organisations (RPOs), empowerment of farmers and results of collective action: introductory note. WDR 2008 Paris Workshop.

World Bank, 2002. Building Institutions for Markets, Washington.

Producer organisations and market chains

Section A

Organisational support for producer organisations

The balancing act of creating a cooperative: the role of outsiders in a Kyrgyz Republic case

Gerrit Holtland

Supporting farmers in developing countries in forming a cooperative requires a two-fold balancing act. Outside support must be balanced with local initiative, while, at the same time, organisational development must go hand-in-hand with business development. Often outside support comes from the most powerful actors in this process, and they are primarily responsible for maintaining these balances. The case study shows that in so doing, cooperatives can become powerful vehicles for economic development and empowerment.

1. Introduction

Although in an ideal world agricultural cooperatives are created by farmers, using their own ideas and resources, in reality outsiders (i.e., non-member stakeholders) often play a decisive role. In the cooperative history of the Netherlands, such outsiders were local leaders (teachers, general practitioners, and priests) and farmer unions who contributed ideas, knowledge and skills. Nowadays, in developing countries outsiders are often development cooperation experts paid by donors or NGOs. In addition to expertise, they often provide financial support. Considering the many difficulties facing farmers in the South, such comprehensive support is justified but it also carries some risks.

Having outsiders support cooperative development has a number of advantages. In addition to providing the cooperative with access to technical and organisational expertise and to capital, outsiders can facilitate sensitive socio-political discussions between farmers and act as brokers between farmers and other, often more powerful, stakeholders. As such, creating a new cooperative in a developing country is an act of balancing local initiative and outside support. At the same time, another crucial balance has to be struck between social inclusion and democratic principles on one hand (in organisational development) and economic, technical and market issues on the other (in business development).

There are also some disadvantages to giving outsiders a major role in setting up a cooperative. Outsiders lack knowledge on local realities, opinions and relations. In addition, large donor budgets and the pressure on projects to produce quick results can easily lead to an over-emphasis on tangible, short-term results to the detriment of aspects that are important in the longer term, such as building up social capital. Moreover, the intervention of outsiders can lead local actors to become insufficiently cost-conscious. Finally, power differences between insiders and outsiders can lead to an unclear allocation of responsibilities and miscommunication. Farmers are tempted to tell donors and project staff what they think donors and staff want to hear, in order to secure (short-term) financial support. Often the final result is that farmers have a low sense of ownership of the cooperative.

This chapter presents a case study of the Naukat Seed Potato Growers Cooperative (NYNOK) in the South of the Kyrgyz Republic. It describes how the above mentioned balances between insiders and outsiders and between organisational development and business development were developed and maintained. Despite a rather short period of support (1999-2002), a robust cooperative of over 200 small farmers was created that not only helps its members earn 25% more income but also provides a large number of non-member farmers with high quality seed potatoes. The chapter presents the development of both organisation and business aspects within the cooperative, gives the main achievements of the cooperative, discusses the role of outsiders, and finishes with a number of lessons learned. We begin by providing a brief description of the context.

2. The context

Naukat is a district situated approximately forty kilometres south of Osh, the capital of the Southern part of the Kyrgyz Republic, which lies at the head of the Fergana Valley. This valley offers fertile lands that can yield bumper harvests if irrigated with water from the surrounding mountain ranges. The valley has been overpopulated for centuries. The city of Osh is among the oldest continuously inhabited places in the world. It has a mixed population of Kyrgyz who are originally nomads whose livelihood depends on grazing horses and sheep at high altitudes, and Uzbeks who are agriculturists and traders. The relations between Kyrgyz and Uzbeks are tense. For instance, in the early 1990s clashes between the two ethnic groups in Osh took hundreds of lives.

Naukat lies at the foot of a mountain range, at mid altitude (at 1200 metres). It combines the advantages of sufficient (irrigation) water, an attractive climate and fairly fertile, accessible land. It is among the most densely populated areas. The most common crops are wheat (65% of the arable area), maize (10%), tobacco (10%) and potatoes (10%). The importance of the latter ranges from 5% in the least densely populated communities to 25% in the most populated communities. At the turn of the century the region's production system was characterised by:
- small unit size (most families have less than 1 ha.);
- low levels of mechanisation;
- low levels of input usage;
- dysfunctional irrigation systems.

The support system is weak. Farmers market their agricultural production either directly to middlemen at the farm gate, or through bazaars in Naukat and Osh. Much agricultural production is finally sold in Uzbekistan. This is particularly true for potato of which 65% is exported via an informal system of independent wholesalers. Credit is very difficult to access due to high interest rates, long and complicated procedures, and strict requirements as to fixed assets due to collateral and credit products that are not well geared towards farmers' needs. At 200 USD per capita, the average yearly income is very low.

In 1997 the Dutch Ministry of Development Cooperation formulated NADPO: the Naukat Agricultural Development Programme. The Dutch wanted to support the Kyrgyz government's agricultural reform program that was much more advanced than those of neighbouring countries in Central Asia. The purpose of NADPO was to improve the institutional framework to facilitate improved production as well as higher profitability for potato (and other crops). The strategy was to create a Private Development Organisation (PDO) that, after phasing out the project, would offer farmers services in the area of marketing, input supply, credit, irrigation, and obtaining technical expertise.

Due to delays in designing a cooperation agreement between the Kyrgyz and Dutch governments, the actual implementation, by Stoas International, started only in April 1999. By then it was too late to work on potato cultivation, so it focused first on improving marketing by storing the potatoes during winter and by setting up a price information system. It also explored different credit options. Parallel to this, work began on creation of the PDO. The founders were to be development NGOs, private farmers' organisations and individual farmers. The assets of the project would be handed over to the PDO, which would use the assets to generate profit for investment in agricultural development.

At the end of 1999 a committee was created to prepare the establishment of a foundation. Members of the committee were representatives of: the regional administration; the Kyrgyz Agricultural Finance Cooperation (KAFC); a private extension service (TES); a public extension service (RADS); and, the Dutch embassy. In addition, each of the Councils of the three villages where the project was to be implemented had representation on the committee.

People were elected or appointed and working groups were created to identify income generating activities. Options were seed potato multiplication, potato storage, advisory and information services, input supply (fertiliser and pesticides), contracting and mechanisation, apple grading, packaging and marketing and, lastly, small scale food processing. In practice, neither the working groups nor the committee ever met. People and organisations were not interested and did not see the potential profitability of most activities.

In the spring of 2000, the trial with potato storage proved to be a non-profitable exercise and the number of options for credit proved to be very limited. NADPO focussed its attention on potato production, yet the imported Dutch Elite seed potatoes were twelve times more expensive than local seed potatoes and farmers were not interested in buying them. So the project had to contract 30 farmers to grow the seed potatoes; the project covered all costs and paid the farmers for their labour. It was at this stage, in the summer of 2000, that the author came to the project and initiated a process that lead to the creation of NYKOK.

3. Organisation development

At this point, a purely farmers' orientation was adopted. Farmers had to have a direct say in all decisions leading to the creation of a Producers' Organisation (PO). To begin with, NADPO staff were trained as to how to communicate with farmers. Second, an inventory was made of all factors and actors affecting the possible creation of a PO. Third, informal meetings were organised at village level to discuss the options for any form of cooperation. Farmers had little, if any, experience with organising themselves, except for traditional communal working parties (*'ashar'*). There were no examples of formal cooperatives that did function.

Yet, farmers proved able to articulate the potential constraints and opportunities for a PO. They pointed out the shortcomings of the village administration and said that the leaders of the PO should not be members of the Village Administration. Leaders should be good, respected and mature people: this meant farmers between 45-55 years old, who were still actively farming (although their sons might do most of the work). The elderly (those over 55 years old) who were normally involved in leadership roles were thought to be too oriented towards village and religious politics and too far away from field realities as they most spent considerable time at the local mosques[1]. People also clearly expressed that they wanted the PO to be composed of three 'independent units' at village level. A very important process result of the informal discussions was that farmers felt they were taken seriously. When the meetings took place, the farmers had already witnessed the good performance of Dutch potatoes in the fields, so they wanted the PO to focus on seed potato production.

In November 2000 these issues were discussed in formal meetings in each of the three villages. Farmers voted to create a PO and per village two temporary leaders were elected based on the selection criteria described above. The mediating role of the project was crucial. As it

---

[1] Naukat was known as a centre for Muslim fundamentalism. The prevailing poverty and the extreme shortage of land offered a fertile ground for recruiting young men for idealistic aims like restoration of the Caliphate of the Fergana Valley.

presented the list of criteria elicited during informal talks and meetings, nobody could be blamed for having encouraged the exclusion of village administrators and (religious) elders.

A draft charter with basic organisational principles was also approved. The key issue was that members elected the board, who would appoint and control the manager, who would in turn appoint his staff. On other organisational issues, farmers were not interested in statutes or formalities. They wanted to talk about practical issues like: how many seed potatoes can each member get? Could they get inputs on credit? How many staff would the PO have and what would their tasks be? The leaders had to think through all of these topics, together with NADPO, as project support was needed on many fronts. The PO did not yet formally exist but in order to create a strong sense of ownership and opportunities for learning, the temporary leaders were empowered to jointly decide on the use of all NADPO resources immediately after their election. The relationships between NADPO and the farmers were shaped in this way, as if a PO already existed.

As a result, the temporary leaders took their responsibilities very seriously. They ensured that the contract farmers for the 2000 season delivered all the seed potatoes they produced to NADPO. Of course this was in their own interest, as the PO would later get the seed potatoes. Next they selected contract farmers for the 2001 season. About 150 farmers were needed to grow the seed potatoes. During the 2001 season they collected membership fees, inspected the fields (to guarantee high quality seed potatoes) and organised credit groups to buy fertilisers and pesticides (most leaders offered their houses as collateral for the group loans from KAFC). NADPO assisted them with training (in field inspections, crop husbandry, storing techniques and marketing) and by providing a guarantee fund for the KAFC loans.

Throughout 2001 the temporary leaders and NADPO settled all legal aspects. This took nearly a year as the legislation was very confusing and nobody had experience with creating a cooperative. Finally, in the autumn of 2001, the Naukat Seed Potato Growers Cooperative (NYKOK) was created. In their first meeting the leaders appointed the crop and storage expert of NADPO as its manager. This was an interesting move as he was an Uzbek, while all members were Kyrgyz. The NADPO co-manager, a Kyrgyz with ambitions to become manager himself, was bypassed. The new NYKOK manager in turn appointed three of the leaders as Village Coordinators to assist him with work at the village level. Next he appointed the chairman, who was a bookkeeper by profession, as bookkeeper of the cooperative. This was frowned upon by some observers ('they appoint each other'), but in this way the hard voluntary work of the elected, temporary leaders was rewarded, their experience was retained and the farmers realised even more that this was indeed their own organisation. With four of the six temporary leaders now employed by NYKOK, the General Assembly elected a new Board, this time a permanent one.

In December 2001 the Kyrgyz Ministry of Agriculture agreed that the project assets would be handed over to NYKOK rather than to a governmental agency (as was initially agreed between the two countries). In January 2002 a contract was signed stipulating that the project would hand over its assets to NYKOK in three stages: in March, July and October 2002 (at the end of the project). In 2002 NYKOK was fully operational. It purchased the seed potatoes NADPO had grown and imported new Elite seed potatoes from the Netherlands. At the end of 2002, NADPO phased out and its assets were handed over to NYKOK.

4. Business development

Discussions as to business development also started in summer 2000. In spring 2000, 54 tons of Dutch seed potatoes had been imported. Yet, these were twelve times more expensive than local seed potatoes and no farmer was willing or able to buy them. Most farmers were not even able to pay for local seed potatoes; often they simply used their own small tubers and many needed a loan to buy seed potatoes for their small plots (usually between 0.1 and 0.2 ha; no farmer had more than half a hectare). For this reason NADPO hired both land and labour from 30 farmers and grew potatoes on its own behalf. Farmers were trained in letting the tubers pre-sprout, planting distances, timing and quantity of fertilisation, and proper use of crop protection. The yield was rather disappointing from a technical perspective: only 22 ton per ha, partly due to poor irrigation practices. Yet, it was sufficient to impress farmers, as local seed potatoes yielded much less (16 ton/ha). So the question became: 'Would it be possible to import such expensive seed potatoes on a commercial base?' A simple economic simulation model was made (using the Excel program) to see how the cost price of seed potatoes would decrease over time. The model proved that after three years of multiplication the costs price would be low enough to make seed potato cultivation profitable.

This raised the question as to how to organise production over these three years. Farmers lacked the money to purchase imported seed potatoes so a cooperative was needed to accumulate the necessary capital. To assist the decision making process a complex economic simulation model was developed that could predict the profitability of both the cooperative and its members, based on assessments of production costs, inputs, yields, market prices and the costs of running a cooperative. The model showed that it was possible to create a viable seed potato cooperative. The cooperative had to multiply seed potatoes in the first season, for itself. This would bring down the cost price for second generation seed potatoes to four times the local seed potato price. Over the next two years these seed potatoes would then be given as in-kind credit to members: in the second year members would repay every kilogram of seed potato with 2.5 kg at harvest time; in the third year with 2 kg. As the cooperative would not incur any costs over those two years, the cost price after three years is roughly the same as for local seed potatoes. Yet, their yield potential would be bigger as would be their market price: roughly fifty percent above the price of local seed potatoes.

Farmers were very enthusiastic about the in-kind credit approach, as it circumvents the need to purchase expensive seed potatoes. The money saved can be invested in fertilisers and pesticides. As the quality of the seed potatoes was very high, yields were much higher than with local seed potatoes. In general the returns to investments on the NYKOK seed potatoes were at least double the return on local seed potatoes. The system solved another major issue as well: how to market expensive quality seed potatoes. As members repaid the cooperative in-kind, they already marketed most of their surplus. And in this way the cooperative was able to guarantee quality (including proper grading), could offer sufficiently large quantities, and had a store in the centre of Naukat town that was always able to satisfy the demand of major clients.

Within this system the long-term profit on imported Elite seed potatoes is shared between the cooperative and its members. It is a precarious balance: if one of the two does not profit enough then the system collapses. The balance can be regulated by the amount of seed potatoes farmers have to repay annually to NYKOK. If the cooperative is in a difficult financial position members have to repay more per kilogram of seed potatoes taken. If the cooperative is doing well, but the profit for members low, then repayment can be reduced.

The simulation model was permanently updated with actual data. It was the basis for all major economic decisions during the project period and even after project completion. NYKOK leadership was trained based on this model. This gave them an in-depth understanding of the long-term underlying economic principles of the cooperative. The model showed that NYKOK was economically viable in the long run, yet to get it going substantial investments were required in fixed assets, especially in stores. NYKOK needed working capital as well, as in the three years that the multiplication system was being established, it would have no income. NADPO provided both assets and working capital over three instalments in 2002, based on clear contracts stipulating the rights and obligations of both parties. The support of the project in the second and third instalments depended on the performance of NYKOK.

In the spring of 2002, NADPO sold the 540 tons of seed potatoes it had produced in the previous two years to NYKOK. NYOK did not have cash, so this was done on credit. NYKOK marketed two thirds and one third was provided as in-kind credit to members. A small amount (the best seed potatoes) was used to multiply on NYKOK's own account. The economic simulation model was used to determine price. The guiding principle was that NYKOK would be able to make some profit, but not so much that it could become careless and lower its financial discipline. The seed potatoes allowed NYKOK to start operating, but the long-term cash flow predictions of the economic simulation model showed that this was still insufficient to finance the cooperative in the first few years after phasing out NADPO. So the project purchased three tractors and a small truck (to haul potatoes from distant fields) for the cooperative on the condition that the vehicles would be leased to private contractors. By extending the lease contract over a period of five years, the cooperative secured a basic income until 2007. In addition, the leasing arrangements reduced the management burden of the cooperative and nullified the risks of (prominent) members defaulting on service payments. Finally, NYKOK needed sufficient storage capacity for at least 1000 tons of seed potatoes. NADPO handed over the two stores it owned, including the necessary ventilators and grading equipment. This was done in autumn 2002 once the cooperative had demonstrated that it had fulfilled all agreed pre-conditions.

5. The results and the impact

The final result of all these efforts is a robust cooperative that serves its members with a number of services such as:

- high quality seed potatoes;
- farming inputs (fertilisers/crop protection chemicals);
- mechanisation services;
- loans (for inputs as well as for other farming activities);
- a quality control system;
- a marketing channel.

NYKOK helps its members earn an additional annual (joint) income of 50,000 USD and to earn itself 100,000 USD. In addition, the clients who buy quality seed potatoes earn an additional (joint) income of about 500,000 USD due to the higher yields they obtain. In fact, NYKOK has transformed the seed potato market. While originally farmers simply used small tubers from the previous harvest as seed potatoes (which were even cheaper than consumption potatoes), now they see the value in selecting proper seed potatoes and readily pay a fifty percent higher price.

NYKOK continues to be robust. After NADPO phased out in 2002, NYKOK continued to import Dutch seed potatoes in 2003 and 2004. In 2005 they tried to reduce costs by importing Dutch seed potatoes that were multiplied in Russia. Yields were reasonable but the quality was still found to be to poor; so in 2006 they reverted back to Elite seed potatoes imported directly from the Netherlands.

NYKOK proved to have gained sufficient social maturity as well and it was able to muster considerable socio-political clout. In 2003 it was recognised as a state seed potato farm, which granted a tax advantage. The state forestry department leased some land at high altitude to NYKOK (see below). On top of that the board and staff managed to convince the three Village Councils to allocate 25 ha of land to NYKOK as well. This meant that its members had more land available for the necessary crop rotation. In return NYKOK assisted the local government when a rural road in the area needed to be renewed.

NYKOK also started to invest; the first profits were invested in a new, large store. The paid contracted work was equally divided between craftsmen from the three villages. Growing member confidence was tested when in 2003 the former NADPO project manager tried to gain control over the cooperative by bribing some members to oust the manager and propose him as a replacement. The attempt failed and in the General Assembly he was openly accused of corruption. At that meeting it was decided that neither civil servants nor anybody known to be (or to have been) corrupt could ever become member.

6. The role of outsiders

Initially NADPO was the driving force behind the creation of NYKOK, but in the end farmers took full responsibility and control. We have already seen that NADPO played a number of roles. It assisted as a mediator in the initial stage when anonymous members indicated in informal talks the type of leaders they wanted. It also developed the economic simulation model that became the basis for most business decisions. But there were other roles as well.

NADPO also was important in providing technical expertise. It had an irrigation component which assisted the Water Use Association with improvements to the irrigation system. In addition, it set up a series of trials on fertilisation, pre-sprouting, use of pesticides, variety trials, storage systems, and irrigation frequencies. Since all this was done via on-farm trials the knowledge generated was immediately available to farmers. The project backed this up with extension services and training courses. Over three years, a system for producing high quality seed potatoes was developed that was well adapted to local circumstances. This complex innovation allowed NYKOK to enter the market with a unique product.

In general, NADPO was very strict as to product quality. At one point it even came into conflict with the leaders of NYKOK. In the autumn of 2001 it was clear that the quality of the seed potatoes was good, but could have been better if the Elite seed potatoes were multiplied at a higher altitude (above 2000 meters) where there are no aphids to transmit viruses. This advice was given to the leaders, but field visits showed that they still intended to plant the seed potatoes on fields near the villages (at 1200 meters). Probing into the reason for this revealed that the leaders did not have land at higher altitudes. NADPO then formulated a clear pre-condition for further support: if NYKOK would not identify fields at higher altitudes then support would stop. Since it was very late in autumn, this was a very tough demand. With great difficulty NYKOK staff and leaders managed to find suitable fields; some were even outside of their own villages. The new fields proved to be very fertile because they had only been used for pasture during the

Soviet era. So in 2002, the seed potatoes were not only virtually virus-free, the average yield had increased to 38 ton/ha as well.

Another role of NADPO was to link farmers to third parties. For example, NADPO had a staff member who organised informal credit groups (for which leaders pledged their assets as collateral) and it placed a guarantee fund at KAFC for these loans. Over the years this system was expanded and the loans could be used for purposes other than potato production. At the end of the project, this staff member was hired by KAFC; initially he was still paid from funds that NADPO had given to KAFC, but after one year he continued as a normal employee of KAFC.

Finally, we return to a purely farmer orientation. Initially the project was supposed to work with government seed inspectors. But they proved to be incompetent and unmotivated. So NADPO decided to train the farmers in field and seed inspection. This meant that farmers were taken to the Netherlands for a thorough training in recognising virus infections found in potato fields. This led to a clash with the vice minister of agriculture (also chairman of the NADPO board) who insisted that he should be part of any study tour to the Netherlands. That both Dutch and Kyrgyz project staff fought to prioritise farmer participation[2] and that they won this battle, was an immense boost to farmer morale.

The training in the Netherlands was indeed taken very seriously by participating farmers. They did not only learn to identify infected plants in a very early stage, they also saw the importance of a proper quality control and certification system. Back home they designed a similar system for NYKOK: field inspectors were appointed and members had to follow their advice. It was quite a struggle to get farmers to accept the fact that they had to remove all infected plants. It was only after the General Assembly in 2002, where members insisted that those who did not follow the advice of field inspectors should be excluded from membership, that compliance was enforced. The expulsion of some members was probably the most significant indicator of member ownership. The leaders were actually hesitant to publicly criticise member non-compliance, but ordinary members compelled them to do so.

There were also some things NADPO did not do. It never entered in any dialogue with government agencies like the tax department or the seed inspection department on behalf of NYKOK. Nor did it secure any

---

[2] They were concerned that having a vice minister in the group would make the programme too formal, and that he would never use the skills acquired.

deals on the market. The cooperative had to find its own way in dealing with socio-political pressure, including corruption. It managed to do so. In 2003 it won a local World Bank tender to supply seed potatoes and it needed to bribe some local officials for this. But the bribe was done with consent of the leaders. And as we have seen above, it successfully used its considerable socio-economic power to strike some interesting deals with the local administration.

7. Lessons learned

At the start of the project farmers did not have any idea as to the value of quality seed potatoes, nor did they have much confidence in the potential value of a modern cooperative. However, in a few years they managed to master some complex innovations and develop a very high level of ownership. The lessons learned are many. Most have already been mentioned. The general lesson is that the many different aspects have to be balanced. Progress in the area of organisational development should be accompanied by progress in business development. And outsiders have to respect the ideas of farmers related to the design of the organisation, while farmers should be open to the technical expertise provided by outsiders.

Although many aspects are important, we can summarise a number of developments and choices that were crucial for the final success of NYKOK:

- The project made very good use of the strong social fabric already in place, especially in the election of the temporary leaders.
- The purely farmers orientation and the permanent and open dialogue between the project and farmers' leaders were clear signals to members that the cooperative was working in their interests.
- A detailed economic simulation model guided internal economic relations in the cooperative and shaped support from NADPO to NYKOK.
- There was a strong focus on quality issues and on training (both on technical matters and on organisational and economic aspects).
- Farmer commitment to NYKOK was fuelled by the huge increase in their income. By becoming members of NYKOK and planting better seed potatoes on 0.15 ha of land, farmers increased their annual family income by 25%! Such a strong incentive can never be generated by merely strengthening bargaining power (e.g., negotiating cheaper inputs or better prices on the market). It was the introduction of a

complex but viable innovation that allowed NYKOK to take such a unique position in the market. Interestingly enough, it seems that the way NYKOK is organised might be the only way to make this innovation viable. Several other projects tried to set up seed potato multiplication schemes in the Kyrgyz republic, but so far none have managed to come up with a sustainable system.

- The fact that the farmers were involved in all activities before the cooperative was formally established provided them with some room for experimentation (both on technical and social issues), rendering the whole exercise more natural. The PO became a formalisation of things that people already knew made a lot of (economic) sense.

Finally, this all boils down to creating trust between outsiders and insiders. In this case NADPO managed to gain the trust of farmers by carefully listening to them, by empowering their elected leaders, by using an economic simulation model as a transparent tool for decision making, and by maintaining very high professional standards in working on technical, economic and social issues.

For the author, working with NYKOK was very challenging. The development of a realistic economic simulation model was, in particular, very exiting. Having seen the positive impact of this model on the decision making process (both from a project and farmer perspective) the method was then successfully applied in Moldova, Albania and Ethiopia as well. Presently these experiences are being used to develop a training course on cooperative development.

Coaching organisational transitions towards an increased market orientation: SNV's experience with producer organisation support in the Bolivian Valleys

Lithzy Flores[3], Dick Commandeur, Giel Ton and Gerda Zijm

This chapter presents an approach towards supporting producer organisations according to their stage of development. The methodology put forward emerged out of 15 years of SNV experience supporting a group of seven smallholder organisations. The authors describe the process and criteria that proved useful for self-assessment within the organisations, emphasizing market orientation, transparency and income generation to reduce donor dependency.

---

[3] Lithzy Flores, Dick Commandeur and Gerda Zijm worked with SNV's PO support program over the period 1999 to 2007. The support programme covered seven POs in the Bolivian Andean Valleys that received support from NOVIB, ICCO and CORDAID. Giel Ton worked, supported by ICCO-PSO, in CIOEC-Bolivia to strengthen the regional and national federation of POs.

1. Introduction

Since the 1990s producer organisations[4] have played a leading role in Bolivia's rural development strategies, both in public policies and in development cooperation by NGOs. Producer organisations managed to profile themselves both as a specific interest group and as an economic sector (Bebbington, 1999; CIOEC, 2000). Many POs received external support during start-up, and their activtities used to reflect the objectives of these external organisations (Swen and Bot, 1999; Commandeur, 1999). At first, activities tended to focus on social and economic assistance for those members who were considered project 'beneficiaries'. After the start-up phase, in particular once external support was phased out, POs were compelled to use an approach that focused on better prices and better markets for their products. Facing the need to readjust to more economic objectives, some POs dissolved while others were able to adapt (Muñoz, 2004).

SNV was closely involved in this transition process of a number of POs. Since 1995, PO strengthening has been central point in SNV's support activities in Bolivia's Andean Valleys. Over the course of more than ten years, SNV developed an innovative approach for coaching POs in their economic activities and business development. This approach has been systematised in various documents. In this chapter we provide an overview of SNV's approach and highlight some central issues that were crucial to the learning and organisational change processes in the partnership between POs and SNV.

The first part of the chapter discusses the most relevant forms of Bolivian producer organizations and their political and legal origins, with an emphasis on the experience of organizations in the Andean region. The second part addresses the development of SNV's coaching approach. This emerged from the concrete experience of long standing support relationships with different types of Bolivian organisations. The chapter presents some key elements of SNV's approach that may prove useful for replication in other countries and contexts.

---

[4] In this text, we use the abbreviation PO for Producers' Organization(s). However, in Bolivia, the more widely used terms are: Economic Peasant Organizations (OECAs) and Producers' Associations.

2. Forms of organization of small producers in Bolivia

Bolivia is a country with considerable organizational experience in the rural areas. For centuries 'ayllu' functioned as the prime political and administrative entity in rural areas of the Andean region. Ayllu were based on membership, regular member consultations, election and rotation of leadership, and are still an important reference in Bolivian rural development discourse. Another important organizational experience in rural areas of Bolivia was the Land Reform process in 1953. Throughout the Andean Valley hundreds of so-called Agrarian Unions (Sindicatos Agrarios) were set up to manage what were formerly 'hacienda' lands. They played (and continue to play) an important role in the management of natural resources and the distribution of use rights for common property, such as land and water. In most communities, the Unions operate as the prime administrative and political authority. In this sense, the Unions are not a primarily market-oriented form of organization[5].

In the 1970s, the Church promoted the creation of multi-purpose service cooperatives involving large numbers of small producers. Although peasant cooperatives were the legal structure of organization that best facilitated combining economic and social objectives, economic results were quite poor and they operated as instruments for channelling social projects rather than enabling joint economic action. Nonetheless, there are still cooperatives, especially in settlement areas of the Yungas and around Santa Cruz, which prospered by adjusting their operations to match market conditions.

At the beginning of the 1980s, when democracy was reinstated in Bolivia, popular organizations such as the Bolivian Union Confederation of Peasant Workers (CSUTCB, *Confederación Sindical de Trabajadores Campesinos de Bolivia*) came to have considerable influence in decision making at the government level and they were able to push through the Supreme Decree that created the CORACAs (*Corporaciones Agropecuarias Campesinas*). These CORACAs enjoyed massive external support over a short period of time, notably receiving donations of tractors for collective use to mechanize their farming activities. Later, in post-inflationary times when structural adjustment policies were applied, most CORACAs

---

[5] The Agrarian Unions may in some places even function as an obstacle to the emergence of other more specialized economic groups within the villages, as these new groups are considered by some to be a threat to Union monopolies (Commandeur, 1999).

were unable to consolidate their services in a sustainable way without that external support.

In recent decades, associations have emerged as the dominant legal construction for POs, promoted by NGOs, international donor agencies and government programs. A civil association is not legally entitled to distribute its assets to members/shareholders (as cooperatives and commercial enterprises do) and therefore turns out to be the preferred legal format for receiving external donations for technical assistance, training, infrastructural investments, etc. Through associations, smallholders are able to start up economic initiatives easily and at low cost. However, once an economic activity reaches a certain level of maturity, associations tend to experience restriction of the legislative framework that curtails their room of manoeuvre (Mendoza and Ton, 2003).

In response to these constraints, some cooperatives and associations chose to organize their economic activities in a separate commercial company. These commercial companies - generally shareholder companies (SA) or limited liability companies (SRL) - emerge in different forms: sometimes they are created as autonomous organizations with a separate management and board; in most cases they emerge as a specialized economic arm of an association or a cooperative. In the latter case, the association or cooperative founds the company and acts as a majority shareholder. Doing so, their legal person is better equipped for accessing markets and attracting commercial credit, but the governance structures continue to overlap with the 'mother organisation'.

3. Characterising the change towards a market-oriented approach

3.1 From receiving benefits to a contributing membership

Over the course of their development as economic organizations, POs realize that if they want to improve economic conditions for their members, they must necessarily identify, adapt and respond to market demands and opportunities. This implies a huge change in thinking. Given that members of many of these organizations expected social services or NGO support, their leaders concentrated on efforts to channel demands externally rather than demanding internal organisational change. However, to survive in the market, an economic organization needs members with sufficient capacity to comply with commercial commitments and contribute products or money to their organisation. To stimulate stronger commitment, the PO has to generate incentives for member-investors and, in many organisations, have been obliged to improve the quality of their membership delisting some of their inactive members. PO leadership has to work to strengthen solidarity amongst members rather than building links with the community in general.

This change has not been easy. Particularly for POs that emerged as an economic arm of the Agrarian Unions, there were leaders (and advisors) who resisted these changes and considered them to be contrary to the 'communal' and 'solidarity' role they wanted their organization to play (Commandeur, 1999). The leaders had to shift from their role of intermediaries with an outward focus seeking projects to implement at the grassroots level, towards a leadership that focussed internally, introducing more demanding criteria in 'commercial' relationships with members. For example, the Association of Men and Women Artisans of Tajzara (AAAT, *Asociación de Artesanas y Artesanos de Tajzara*) developed a niche marketing strategy with higher quality requirements for textiles produced by members. However, many members had become associates because they were after cheap raw materials, not to develop a joint marketing initiative. Due to more demanding policies, they decided to leave the association. For AAAT this shrinking membership was a healthy development that facilitated commercial success as an economic organisation.

3.2 From a product-driven to a demand-driven approach

In the past, POs had a 'product-driven' approach. They argued: 'our product is good and the organization must sell it at the best possible price'. However, market realities forced them to change their position. The profits from non-processed agricultural commodities declined due to the increasing import (both legal and through smuggling) of cheap agricultural products from Argentina and Chile (Prudencio and Ton, 2004). As a consequence, the sustainability of various POs was at stake. The POs that centred on input provision and technical assistance for production (for products such as beer barley, milk and wheat) entered a crisis because the price fell and preference was given to the massive purchase of standard quality products from abroad, rather than products of varying quality from local markets.

Nonetheless, some economic activities that are not at all profitable for the PO as an organisation proved to be positive for the economic development of members. This provided a rationale for continuing external support to those POs that for example, regulated product price in the intervention area with their purchasing capacity, like CORACA-Aiquile. Even if they bought a mere 10% of production in any given area, simply their presence as possible buyer forced intermediaries who wanted to buy in these regions to pay at least the price fixed by the PO. This PO function could only be maintained with support from donor agencies, which resulted in dependence and tended to reinforce the POs' identity and discourse regarding their social role in contrast to their commercial role.

Most POs, however, responded to market cues quickly and understood that in order to survive as a viable economic actor, they clearly needed to change from a 'product-driven' approach to a 'demand-driven' approach: POs had to adapt their crops and product quality to match market demands. This resulted in basically two economic strategies:

- Improve prices by creating added value through (basic) processing.
- focus production on the strict demands of market niches, both within the country (organic market, supermarkets, tourists) and at the international level (fair trade, organic trade).

As PO businesses had to benefit members and at the same time generate a surplus for ensuring operation of the organization, the POs started to prepare 'business plans' as a new management tool, instead of or in addition to the 'annual operational plans' generally required by donors.

One example of this change is the Association of Small Wheat Producers (APT, *Asociación de Pequeños Productores de Trigo*) in Chuquisaca, which was set up around wheat production, but which now promotes organic production and has been successful in commercializing non-traditional products such as amaranth and beans. Although these crops have not replaced wheat production (mainly for self-consumption), they have become priority crops for APT marketing and a cash crop for members.

3.3 Changing logics in staff recruitment

Many POs hired technical personnel through projects or NGOs, which generally focused on topics or approaches prioritized by the international donor agency. Typically, donors supported POs that had staff for technical support in production, for training and for organizational development, as well as for promoting gender equality and the participation of women.

With waning donor support, POs sought cheaper personnel to replace technicians from the support organizations. Some organizations only hired technicians who were members or who came from the same village or region, or at least with a peasant background (Ruralter, 2005). This requirement tended to conflict with other requirements, for instance professional quality for performing the assigned tasks. An example was in APT Chuquisaca where it took time to convince the leaders of the need to hire a professional in commercialisation that did not come from a peasant background. After they gave this person a try and experienced the positive economic returns of his activities, this same person was appointed some years later as the general manager of the organization.

There is a certain tension between the need for qualified professional technicians to start a successful commercial activity and the difficulties most POs face when trying to generate sufficient resources (a 'margin') to pay for technicians' salaries during the kick-off phase. This tension can be solved through external support from the state or international donors so that, at least initially, POs can hire technicians and extension workers but within a framework of gradual phase-out of external support (Ton and Jansen 2007). However, there is another tension within the PO: the leaders (generally from poor households) must define the salary of a professional who will work for them; but this person will earn much more than the leaders ever will. PO sector salary scales continue to be quite low and are not very attractive for qualified professionals.

A relatively new development is the demand by several POs for more specialized technical assistants who will not necessarily become employees. These POs require a consultancy-type of support related to their need to resolve bottlenecks in product innovation, technology, marketing, diversification and other activities considered priorities for the development of their competitive advantage on the market.

4. Guidance and advice from SNV

At the beginning of the 1990s, SNV Bolivia decided to focus its work on grassroots organizations. For the Andes area, this crystallized in a program for strengthening the economic organizations of smallholders with a combination of financial support and organisational advice.

The work methodology was implemented as of 1995, passing through different phases and kinds of work.

4.1 Coaching according to development phases

In the beginning, SNV's approach to strengthening POs, was almost always related to productive investments to free the production potential of the region. In this setting, investments were made, for example, in the production of potatoes, wheat, wool textiles and maize. The aim was to try to find an income generating role for the PO: mainly the production of seed and commercialization. SNV technical assistance developed a general training programme for organisational strengthening: planning the organizational activities and ensuring a democratic and transparent structure to enable active member participation in decision-making (Flores *et al.*, 1999; Peters *et al.*, 2000).

However, experiences with different producer organisations and different contexts provided the elements to develop a new, more flexible and coherent methodology for the organisational strengthening of POs (Swen and Bot, 1999; Toornstra 2000). A core element of SNV's new methodology to support producer organisations towards becoming dynamic chain players was recognition of the need to differentiate between the various developmental phases POs undergo. The following three phases were identified, each with its own activities and indicators: start-up, strengthening and consolidation (Toornstra, 2000). This methodology formed the first real framework for SNV's PO support.

SNV's technical assistants became increasingly concentrated on management issues and business development, rather than technical training. Intense processes were undertaken for: leadership building and training; promotion of women's participation; and the development of normative instruments and mechanisms for organizational operations. Considerable progress was made in terms of increasing awareness among producers as to the importance of their organization, clarification of the process for PO capacity building, and the participation of women producers (Dulón, 2001). The achievements were widely recognized in Bolivia and enabled SNV to profile itself as an expert in PO strengthening.

4.2 Self-assessment of organizational and economic performance

The majority of the POs were not financially sustainable as they generated insufficient resources from commercial activities. Therefore, SNV started to pay more attention to the organization's commercial performance. It became clear that both managers and leaders of the POs lacked sufficient information for making sound decisions related to the organization's economic activities. General information was available, but no details as to the turnover, margins, profits or losses of each activity. These details were needed in order to make decisions related to concrete business activities. For example, CORACA Aiquile had to decide whether it was made more sense to provide transportation services using the organization's truck or whether they should outsource this service; APT Chuquisaca needed information as to the cost of commercializing different collected products; ADEPLECH needed to know the cost of managing its own concentrates plant.

Together with the managers and leaders, SNV developed instruments to register information for each PO in relation to the indicators and information most relevant for them ('*La Carpeta Roja*'), principally on four topics: self-management, income, external relations and business management (Zijm 2005; see annex). The data were registered in the same format for each organization to facilitate sharing and learning between POs. This exchange of information and discussion on key issues gave rise to collaboration. The data on every PO were subject to participatory analysis in a learning platform made up with the leaders and managers of the seven involved POs. An important instrument in this process was the external audit, which did more than just monitor bookkeeping; the external audit was used to generate information for board members on management and administration systems, and to generate transparency in the financial management of POs.

This self-assessment process was directly related to a project funded by ICCO, NOVIB and CORDAID. The POs had to provide the required information and show progress achieved during that period. SNV had a process-enabling role: the POs themselves convened and presided over the evaluation meetings. SNV was able to develop a monitoring system, which was sufficient for accountability to the funding agencies and at the same time was so concrete and transparent that it became an essential element in the learning process for the POs. As a result of this process, SNV developed a self-assessment manual for POs (Zijm, 2005). The central features in the design of this self-assessment system are sumarized in the following text box.:

M&E Design feature	Expected Outcome
The evaluation meetings were convened and chaired by the POs themselves (on a rotational basis).	Ownership of the evaluation and results.
The following players were invited to the meetings: the manager and the president of the PO. Both the managers and presidents of the different POs exchanged experiences with one another.	Space to work on the relationship between management and board. The exchanges encouraged a mutual 'from organization to organization' learning process between peers.
The meetings were held on a regular basis (every two months) over a three-year period. There was little variation in the group of participants. They got to know one another very well, both at the personal level and at the level of the PO.	A mutual relationship of trust was created, which often gave rise to mutual assistance. There was social pressure within the group to insist on the provision of real, concrete and correct information.
In every meeting, the participants were supposed to bring information related to questions prepared by the organizing PO. They also analyzed and discussed annual reports in these meetings.	Evaluation based on real and concrete data.
For the annual report, the organizations used a uniform format for submitting information. The reports were presented in a simple way, using histograms and graphs.	Comparable information; the organizations could measure their performance amongst themselves. Leaders and managers learned about how they could present information in an understandable way.
The exchanged information was related to institutional information about the PO, which was not limited to program funding. The leaders were able to make a complete financial analysis of their organization. Before, generally the financial information presented was partial and biased.	Increased control by the leaders over personnel. A better perspective of the organization's (in)efficiencies in relation to the other POs ('benchmarking').

4.3 Organization of services at the departmental level

With the purpose of institutionalizing the learning and cooperation between the producers' organizations, SNV promoted and guided the creation of regional and national platforms and federations, whose role was to represent the sector and provide specialized services, explicitly recognizing the characteristics and needs of the POs. In two departments (Chuquisaca and Tarija), PO Chambers were set up, which was a fundamental step towards official acknowledgement by the public and private sectors. The chambers were directly linked to a national central organization, namely the Coordinator for Integration of Economic Farmers' Organizations (CIOEC Bolivia, *Coordinadora de Integración de Organizaciones Económicas Campesinas de Bolivia*), which in turn enhanced the creation of similar regional organizations in other departments.

For these regional organizations, representing the sector was and continues to be, their most dynamic and successful role. The provision of effective support services to POs in the department is still the biggest challenge. SNV developed a methodology for preparing business plans in accordance with the particularities of an association's undertakings, with the intention that these regional organizations would provide this as a service to their member organisations. SNV felt however that even though further progress was made in the provision of specialized services by the departmental and national level organizations, both the quality and quantity of services that can be rendered by the regional organisations fall below expectations.

4.4 Linking with other chain actors and service providers

Because of the weaknesses detected in service provision to POs through the regional and national federation of CIOEC, SNV decided to look for alternatives amongst local actors for fulfilling this role. Thus, as of 2004, SNV began developing advisory services for a much wider range of organizations and institutions, aimed at strengthening the POs. PO demands for support in the development of commercial activities were met by partnering with private sector entities offering more advanced commercial support services, for example the affiliation of APT Chuquisaca with the Chamber of Industry and Commerce.

At the national level, space for presenting sectoral demands increasingly opened up with the government (CIOEC, 2000). The state defined that Local Economic Development and Productive Chains were the central axes of the national Agriculture and Livestock Development Strategy in 2004, whereby POs were recognized as privileged actors for organizing and planning development of the productive chains. This political acknowledgment of their role in rural economic development created room for manoeuvre in the departments for discussing value chains problems together with other stakeholders. Two interesting examples are the Association of Artisans - Inca Pallay - that started to play an important role in the Tourism Table in Sucre, and CIOEC Chuquisaca, which took charge of coordinating the Agriculture and Livestock Table of the Competitiveness Council of Chuquisaca.

SNV built on its experience in other programmes with the public sector - first at the municipal and then at the departmental level. As public entities were more and more motivated to support the productive sector, SNV's experience of working directly with POs greatly facilitated its capacity to assist public, private and academic institutions with expert advise. SNV focused on training technicians from municipalities, prefectures, NGOs and development cooperation programs, to elaborate business plans for associative companies; it also assisted the Departmental Competitiveness Councils (with public, private and academic actors) to support productive clusters; SNV also worked with municipalities to promote the development of School Breakfasts based on local products, which POs could provide.

Emphasis on the role of supporting institutions in the PO environment was aligned with the recent corporate strategy of SNV: to focus actions on capacity strengthening of meso-level organizations, and the provision of paid advisory services. In a way, this has reduced the special focus on POs in SNV Bolivia. Nonetheless, the learning and methodologies developed over the past 15 years in the PO strengthening programme are still a main source for products that SNV is offering to public and private meso-organisations in Bolivia to support their efforts to strengthen producers' organizations with the final aim of poverty alleviation.

References

Bebbington, A., 1999. Capitals and Capabilities: a framework for analyzing peasant viability, rural livelihoods and poverty, World Development 27(12), 1999.

CIOEC, 2000. Agenda para el Desarrollo Estratégico de las Organizaciones Económicas Campesinas, CIOEC-Bolivia, La Paz.

Commandeur, D., 1999. Las Organizaciones Económicas Campesinas en su contexto, SNV, Bolivia.

Dulon, R., 2001. Estrategia de Género para Organizaciones Económicas Campesinas.

Flores, L. *et al.*, 1999. Desarrollando Capacidades para la Participación; 7 módulos de capacitación a promotores, SNV, Sucre.

Mendoza M. and G. Ton, 2003. Procedimientos Jurídicos y Tributarios para Organizaciones Económicas Campesinas, CIOEC-SNV, Sucre.

Muñoz, D., 2004. Small Farmers Economic Organisations and Public Policies; a comparative study, Plural Editores, La Paz.

Peters, C. *et al.*, 2000. Desafiando el Futuro: curso básico para dirigentes de las OECAs, SNV, Sucre

Prudencio, J. and G. Ton, 2004. Integración Regional y Producción Campesina: la urgencia de políticas de soberanía alimentaria, CIOEC-Bolivia, La Paz.

RURALTER, 2005. Factores de Éxito de Empresas Agrícolas Rurales, RURALTER. Quito.

Swen, H. and F. Bot, 1999. De tractores y auditores, SNV, Sucre.

Ton, G. and D. Jansen, 2007. Farmers' Organisations and Contracted R&D services: service provision and innovation in the coffee chain. Markets, Trade and Sustainable Rural Development Strategy & Policy Brief #3, WUR, Wageningen.

Toornstra, C., 2000. Buscando una Identidad Empresarial Asociativa, SNV, Sucre.

Zijm, G.M., 2005. Manual para la Auto-evaluación de las Organizaciones Económicas Campesinas, SNV, Sucre.

Annex. Development phases of producers' organizations (Zijm, 2005)

	Start-up phase	Strengthening phase	Consolidation phase
Self-management	• Operational rules embedded in normative instruments such as by-laws and regulations • Interested members • Openness to female participation in the organization's principal activities • A partially participatory structure	• A clear vision and mission defined in a participatory manner • Clear and representative structure • The Board composition considers portfolios based on the organization's purpose • The organization has a strategic plan • Defined membership • Participation of women • A structure that offers equal opportunities to men and women in economic profit-sharing • Efficient use of human resources • Trained promoters	• The Board defines policies and exercises control over them • Committed membership • Problem-solving capacity • Clear and representative structure between different levels • Effective internal control • Participation of women in decision-making • Separation of activities according to their function • The actions for implementation are defined in the implementing agency
Business management	• Simple and transparent accounting • Audit comments solved • Feasibility studies undertaken for economic activities	• The Board defines annual commercial goals • There are financial reports per economic activity, which are managed by the Board • Financial statements per activity • Credit management system • Services are provided in a profitable manner • External audits with positive opinions • Economic activities take adequate natural resources management into consideration • Market studies help to define commercial strategies • Business plans are the basis for taking business decisions	• Entrepreneurial administrative - accounting system • Audits with positive opinions • Definition of the PO's equity • Credit management system in operation • The organization manages a variety of internal control instruments (reports by economic activity, commercial plans, database with commercial relations,…) • Strategies to cope with the commercialization risk • Clear delegation of functions between the Board and hired personnel • External institutional audits with positive opinions

	Start-up phase	Strengthening phase	Consolidation phase
Own income	• Producers obtain income from associative economic activities • Rules for their own contributions	• Direct and indirect cash contributions to the organization • Profits from commercial activities • 25% of the fixed and recurrent expenses are paid with the PO's own income • Clear differentiation between commercial (profit-yielding) and non-commercial functions	• Fixed and recurrent expenses are 100% covered • The direct and indirect monetary contributions cover 25% of fixed and recurrent expenses • Increasing net income
External relations	• Relations with other entities and municipalities	• Institutional relations with territorial organizations, NGOs and the Municipal Government • An active member in a higher-level representative organization • Participation in projects • Participation in the productive chain of its principal activity • Development of commercial relations	• Agreements with other entities and municipalities • Actively promotes functioning of the organization, with commitment • Negotiation capacity • External fund-raising • Establishment of horizontal and vertical relations • Agreements for technical assistance and other services with indirect stakeholders in the chain • Trade agreements with other direct stakeholders in the chain

Producer organisations and market chains

Creating a balance between loyalty and efficiency: the importance of organisational culture for the market integration of coffee producer organisations in Chiapas, Mexico

Olga van der Valk[6]

This chapter examines how local culture can be a decisive element in organisational processes of producer organisations trying to link effectively to international commodity chains. A case study of coffee growing organisations in Mexico serves as an example.

1. Introduction

Integration into market chains is currently analyzed by researchers, and policy and development agents as an alternative pro-poor strategy for

---

[6] The author worked from 1990 until 1999 as an export marketing director for the coffee cooperative Unión de la Selva. In 1999 she was cofounder of the consultancy KAFFE (Konsultoría Administrativa, Financiera, Fiscal y Empresarial), where she worked until her return to the Netherlands in 2003. In 2000 the groups working with KAFFE joined commercial operations and established the farmer owned export company Comercializadora Más Café.

the economic and social development of small producers worldwide. In this process of market integration, market mechanisms impose specific conditions on small farmers. In view of the growing power concentration of multinational corporations and retail organisations, producer organisations are considered a viable option to allow small farmers to integrate into market chains and fulfil market conditions of scale, volume and quality, while maintaining some autonomy and bargaining power.

This chapter focuses on the importance of organisational culture when introducing changes in group functioning to facilitate market integration. The organisational culture of the group forms an implicit framework of codes and values determining how group members (should) behave and act. It reflects belief systems and shared values and is shaped by national culture, history and interaction with other social structures in society. Interventions will not be sustainable if not linked to this deeper level of group identity and shared understanding of how the world works.

This chapter is structured as followed: after defining the concept of organisational culture, an historical overview of peasant movements in South Mexico is provided, concluding with an analysis of the main components of their organisational culture.

Then two specific coffee producer organisations, the Unión de la Selva and the Comercializadora Más Café will be introduced. For both organisations, the influence of introduced changes on organisational culture will be compared.

2. Organisational culture

Definitions of organisational culture (Parker, 2002; McNamara, 2005) mostly come from sociology and organisation management studies. Schein (2000) considers organisational culture to be: *'a pattern of shared basic assumptions that the group learned as it solved its problems of external adaptation and internal integration, that has worked well enough to be considered valid and, therefore, to be taught to new members as the correct way you perceive, think, and feel in relation to those problems.'*

Thus formulated, organisational culture reflects the identity of the group and its experiences in the mobilization of *internal resources* when interrelating with the *external* political *environment* (Honda, 1996). This dynamic can also be distinguished in Chiapas: the organisational culture of current producer organisations was formed by their history as a social group, and their history of organising and interacting with their environment, which for the Mexican movement in particular has meant dealing with the State. In continuation, social identity and grass root community cohesion will be described as factors of internal integration. Then external adaptation of farmer organisations will be described from a historical perspective with the formation of clientelism relations and assimilation of external agents.

3. The roots: identity and community decision making processes

Before they established visible regional organisations in the 1970s, indigenous communities in Chiapas had strong institutions[7] founded on a feeling of belonging to a community that was consolidated through various forms of solidarity and consensual agreement. The collective voice in community and regional assemblies to this day reflects more than just an agreement: it denotes belonging to a social and political whole (Leyva, 2001). At community assemblies, matters are discussed at length and all participants have the right to fully express themselves. The assembly passes authority and power to the representatives, whose exercise of power is as much an obligation as a responsibility and a privilege to 'serve the community'.

Regional low intensity warfare following the Zapatista uprising in 1994 and conflicting political affiliations have created tensions among Mexican social grassroots movements since the 1990s (Leyva, 2001, Legorreta, 1997). Nevertheless, the farmer organisations acknowledge a shared identity that is connected to ethnicity, a history of suppression and a way of life at subsistence level. Farmers address each other with the title 'compañero' (comrade), indigenous people use the (Mayan) word 'Kaxlan' as a generic title for non-indigenous people, and certain references to their position in society are recurrent in conversation: 'somos pobres' (we are poor) (Mattiace, 2001).

---

[7] Leyva refers to the "comon" in the region of the Cañadas, with a twofold meaning: that of the community assembly, and that of the collective voice, the consensus created in assemblies. Mattiace explains that through more regional "exposure", the Tojolabal indigenous from Las Margaritas came to identify early on (1940s) with localized *ejido* communities.

4. Clientelism defining the relationship between the State and grassroots organisations

Land reform[8] did not take place in Chiapas as it did in the 1930s throughout the rest of Mexico. To evade affecting the interests of local elites, the Federal Government induced landless people to colonize the virgin forests. The subsequent struggle for land rights was not the only impetus for small farmers to organise themselves into regional unions; the Catholic Church's also catalysed this process with the promotion of social emancipation through liberation theology[9]. It was during preparation for the first Indigenous Congress in 1974, organised by the Catholic Church, that community leaders had the opportunity to cross village borders to discuss their marginalized position in Mexican society (Legorreta, 1998).

On the national political level, the one-party regime ruled for more than 70 years[10]. Throughout these years, PRI consolidated its stronghold of authoritarian clientelism, which can be defined as a relationship based on political subordination in exchange for social entitlements and rewards through corporate party institutions (Fox, 1994, Van Osten, 2006). Party-affiliated mass organisations put dissident organisations at a serious disadvantage as the conduits for material and organisational benefits (Van Osten, 2006).[11]

In the 1980s, a new generation of politicians aimed at reforming the party-corporate structure by establishing a direct link between the president and the (electoral) population. New committees were set up in the local communities for the implementation of national social programmes. The committees channelled subsidies and inversions,

---

[8] Under president Lázaro Cárdenas (1934-1940), land reform took place with the formation of eijdos: a form of communal property connected to a settlement, with usufruct assigned by the state.

[9] The pastoral agents promoted the revalorization of the indigenous identity, to the point that people started to believe that all injustice comes from the outside. Legorreta (1998) comments on recurrent arguments put forward by indigenous leaders saying: "what we do wrong is not our fault, because these customs were not ours, all wrong has come from the outside world and we have learned from the Spanish".

[10] The Partido Revolucionario Institucional (PRI) ruled from 1929 to 2000, when the liberal conservative Partido Acción Nacional (PAN) won the presidential elections. The elections in 1988 were characterized by electoral fraud and were assumed to have been won by the leftwing opposition Partido de la Revolución Democrática (PRD)

[11] In Chiapas, it resulted in indigenous voting patterns that had little to do with supporting political alternatives for the future, but rather considered the vote to be a resource for the here and now: finishing a road, building a school, the small benefits that help to solve ancestral problems which shape daily live (Fox 1994).

applied to cushion the effects of neo-liberal politics of austerity for the poor (van Osten, 2006).[12] Even though most of the resources from those programmes were delivered through semi-clientelistic channels, they provided openings for autonomous organisations in society to exercise civil rights without relinquishing political rights (Fox, 1994).[13]

5. Political brokers as an intrinsic part of farmer movements

External adaptation of farmer organisations was also facilitated by means of political brokerage. Political brokers played a key role in mediating state-society relations (Fox, 1994; Harvey, 1998). In time, traditional patronage that aimed at integrating social groups in state-controlled corporate structures was replaced by a new bargaining style that bypassed those structures. The State started to recognize autonomous movement leaders as legitimate interlocutors as long as they steered clear of overt political opposition.

This development was accompanied by a new style of political brokers, who since the 1970s had come to affiliate themselves with grassroots organisations, and were mostly university educated external agents: left wing agronomists, priests and Maoist socialists (Leyva, 2001). Their presence as advisors[14] in farmer organisations helped to strengthen regional and national movements, linking communities directly with federal agencies, and bypassing traditional political elites at municipality and state levels.

The first generation of advisors generally started their work in the farmer organisations with scarce financial support and close to the community life. They respected the authority of the local communities and affiliated with their natural farmer leaders. In return, they were

---

[12] SOLIDARIDAD, the national solidarity program, promoted by Salinas de Gortari after the 1988 elections, allocated disproportionate amounts of resources, mainly for public works, to newly established committees in areas of strong centre-left opposition and centralized power by bypassing traditional political bosses and the opposition.

[13] A leader of the Union of Unions commented in 1983: "Though we tried to establish a relationship of dialogue with the state governments, this was not possible. For them there were only two kinds of organizations: we were either in favor or against them. Much as we tried to tell them we were neither, that we wanted a new relationship, they did not accept it." (Legorreta, 1998, p. 93).

[14] The term "advisor" (*asesor* in Spanish) refers to a person who is not a member but neither belongs to the staff (employees). In the organizational chart his or her function is commonly not placed in the line of command, but as an additional "advisory board". As farmers turn to the *asesor* for advice and explanations, the function is more political than entrepreneurial, even though *asesores* may directly interfere in commercial operations.

rewarded with trust and a status of respect and authority within the communities, and received discretionary financial and decision making powers from the farmers.

By demonstrating negotiating skills and strategic vision, the advisors enlarged the network of these organisations and were able to define the strategies of the organisations regarding its relation with the outside world.

As spokes(wo)men they came to symbolize the identity of the organisation, though never assimilated its ethnic elements. Both farmer leaders[15] and advisors used their personal networks and political skills to translate the 'collective voice' of the communities to negotiate civil and social rights as well as material benefits.

Within farmer organisations, due to their political character and difficult relations with their external environment, internal political struggles were a recurring theme (Harvey, 1996). In some cases this would lead to marginalization or expulsion of staff or advisors (Harvey, 1996) and occasionally to repudiation of the departing faction[16]. Sometimes it would lead to the splintering of organisations.

In conclusion, the organisational culture of producer organisations in Chiapas is formed by their history as a marginalized social group in constant political negotiation with their surroundings, in particular the state apparatus. It is also formed by strong traditions of decision making by consensus as social and ethnic groups. The groups maintain a tradition of representation based on authority. Political negotiation skills are valued for tradeoffs with a hostile environment, with farmers trading off political rights for direct benefits. The incorporation of external agents (advisors) influenced organisational functioning and strategies. They were successful in strengthening the organisations' position in a changing clientelistic relationship with the State. The relationship between advisors and farmers was marked by trust and loyalty and a low level of accountability, representing a new form of paternalism.

---

[15] The term "*dirigente*" for the stronger village and organization leader has a connotation of (internal) authority and (external) influence, with characteristics of a political broker. Advisors and farmers shared the terminology of social emancipation, like the "the peasant struggle" (la *lucha campesina*)

[16] Douglas (1995) describes how the identification of "scapegoats" in a group functions as a mechanism to safeguard group cohesion and identity. Honda (1997) calls these processes of reality interpretation and blame attribution "framing processes", one of the analytical units of social movement research.

6. Linking roots to markets: a case study in Chiapas

At the turn of the century, the Unión de la Selva and the groups that would form Comercializadora Más Café, shared the traits of organisational culture described for the farmer movements in Chiapas, though their commercial experience as independent marketing organisations differed.

The Unión de la Selva was officially founded in 1979, but did not export coffee until 1988. In subsequent years, it built up an international market position until 1998, when due to financial problems it became subject to a restructuring process, after which it lost most of its international market.

The six organisations that founded the Comercializadora Más Café in 2000 had only just begun to coordinate commercial activities in October 1999. Nevertheless, several of them had previously commercialized through the Unión de la Selva. The process of commercial coordination was facilitated by KAFFE. The firm introduced management tools and organisation charts in each organisation and a training programme for the six organisations together.

The Unión de la Selva reacted differently to newly introduced administrative management systems than the organisations associated with Comercializadora Más Café did. In the following section both organisations will be described and compared.

7. Unión de Ejidos de la Selva

The producers' organisation Unión de la Selva[17] began as part of several historic socio-political grassroots' movements in Chiapas active since the 1970s. The Unión was a pioneer in developing an economic strategy somewhat different from the other organisations by starting with a focus on productive and commercial strategies before other regional organisations did (Harvey, 1998). Its main advisor, an agronomist from the University of Chapingo[18], in the early 1980s supported the

---

[17] The Unión de Ejidos de la Selva commercialized coffee through one of its legal structures, the Union de la Selva, Sociedad Civil.
[18] The main advisor has worked with the Unión de la Selva since the early 1980s. Particularly during the first 15 years working as an advisor with the Unión, in the development of the organizational strategy he successfully teamed up with the indigenous farmer leader of one of the founding communities of the Unión, who during those years acted as president.

three founding communities[19] of the Unión in setting out an economic strategy in which the negotiation with local and national authorities was as important as linking up with the international market. An important feat was the successful purchase of a coffee processing plant[20], in which the advisor played a key role in bringing together peasant movements of different political profiles[21].

In spite of an initially bad experience due to the non-payment of a container in 1988, which resulted in one-third of the membership deserting, in 1990 the organisation made their first shipment to the European fair trade market. Starting from scratch, with no telephone at the office and a fax machine at the local parish, in the following eight years the organisation managed to build up a marketing position in organic, fair trade and quality niche markets, mainly in Europe. Soft loans and subsidies from national pro-poor government programmes and NGOs[22] generated an enterprise boom. Investments were made in processing equipment, a coffee roasting plant, a 40-ton truck, four coffee houses in Chiapas and Mexico City, and several community programmes. The apparent commercial success came to a halt in 1998, after a bad year due to declining international coffee prices.

The non-delivery of a coffee contract and failure to return the received prepayment to a European buyer, not only exposed severe structural financial problems[23], but also exposed how the Unión was

---

[19] The three communities are located in the Cañadas in the municipality of Las Margaritas: Cruz del Rosario, Nuevo Momón y Nuevo Monte Cristo. Key members of these communities to date maintain a strong (informal) influence on the decision taking of the Unión (González, 2001)

[20] The Unión de Productores de Café de la Frontera Sur (Union of Coffee Producers of the Southern Border), was formed in April 1990 as an alliance with local affiliates of the independent CIOAC, the state party affiliated CNC and the Teacher-Peasant Solidarity Movement SOCAMA. According to Harvey (1998), this convergence of independent and official organizations around economic concerns represented the new type of peasant movement promoted by reformers within the state and UNORCA (non party affiliated peasant movement).

[21] This success resulted in the governor of Chiapas forcing the regional director of the National Indigenous Institute (INI) to resign, because of having supported the purchase by providing access to funds (Harvey, 1998).

[22] Larger financiers were the governmental programme FONAES (Fondo Nacional de Apoyo a Empresas en Solidaridad) and the Inter American Foundation (IAF). A programme directed at the alleviation of poverty through social enterprises, FONAES co-invested in many projects by the Unión: several shops of "La Selva Café" received finance (for infrastructure, purchase of a shop window coffee roaster); a 30 ton truck; a roasting and packing plant; a health centre for women; and, small community initiatives (bakeries and sowing ateliers).

[23] Besides the debt with a European bank, the Unión had to restructure a debt with a Mexican bank.

functioning. The Unión had become a mixture of separate private and social enterprises with cash flows and decision making procedures that were not clear to its members[24]. The structure of the umbrella Unión, did not guarantee proportional voting rights or property rights to its associated members. Founding communities had a strong influence on general strategies, while communities in other regions who had affiliated later and contributed most coffee, were rarely elected in the representative bodies of the Unión. This way of functioning was accepted because communities prioritized access to the fair trade market with higher prices, over their rights to a proportional say and representation in the organisation.

An external auditor intervened in the organisation, financed and monitored by European market partners, a bank and a fair trade certifier. In the export activities, in one year he managed to set up the necessary management control systems and strict budgetary control that had been lacking[25], but could not change the organisational functioning of the Unión[26]. As a result, the Unión paid off its debts to the European buyer,

---

[24] Though opinions on the causes of the crisis differ, the Unión probably was victim of its own reputation of success, as each newly started business generated a cycle of new business and project opportunities. This would give the Unión more visibility and reputation as a successful organization, new opportunities from state programs etc. Entering into new commercial activities unfamiliar to the Unión generated high start-up and learning costs, particularly in the case of the first coffee shops La Selva Café established in Mexico City, San Cristóbal, and Mataró (Spain); and on installing the infrastructure of the roasting plant Tenam in Comitán. More than deficient administrative systems, at the end of the 1990s the Unión lacked sufficiently skilled personnel and sound feasibility studies for all projects except exports. In addition to the organization's lack of definition regarding the complexity of enterprises, overall coordination and planning of different investments in relation to expected returns on investment (ROI) and expected cash flow, was lacking. Most of all the Unión lacked internal accountability in which those in charge and taking decisions also assumed responsibility towards board and members.

[25] The auditor, in close collaboration with the board of representatives and advisor of the Unión, set up a restructuring plan regarding green coffee exports. After its approval by the Unión, export operations started in October 1998. The plan included shutting down all financial flows going from the commercialization of green coffee to other projects under the umbrella of the Unión; a high turnover rate in stocks and financial resources and agreements on putting a part of coffee earnings towards the repayment of debts. Monthly balance sheets were made and the board was kept informed of progress. As a result, the debt to the coffee buyer in Europe was fully paid off in one coffee season (June 1999).

[26] The failure of the Unión to comply with binding conditions set by FLO regarding internal democracy lead in 1999, to the temporary suspension of the Unión from the fair trade register (de-certification). Suspension was related to the failure to formally and effectively regulate how members obtain(ed) (property and voting) rights by participating in the economic activities of the Unión and was effectuated after the set period for corrective actions had expired.

but lost credibility with its European market partners, who could not be assured that the same problems would not arise again in the future.

On the other hand, during that year of restructuring, the head of the marketing department[27] who was responsible for the implementation of austerity in financial and commercial management systems, was gaining credibility with market chain partners. An internal political struggle followed[28], and led to the departure of five employees in June 1999, to establish the consultancy firm KAFFE and start implementing similar management systems in the six organisations, leading to the establishment of Comercializadora Más Café a year later.

8. Using the same tools, but with different results

Thus, in the Unión de la Selva and in the associated groups of Comercializadora Más Café, the same financial-administrative and internal control systems were introduced. Nevertheless, in the first case (Unión) this did not lead to changes in organisational functioning, and finally led to a decline in international market position. The associated groups working with KAFFE on the other hand proved very open to the proposed interventions and became commercially successful. How can this be related to the organisational culture of both groups of associated farmers? As culture was defined in relation to external adaptation and internal integration, these will be compared for both cases.

9. External adaptation

During the 1990s, coffee prices dropped heavily as a result of the abolition of the quota system by the International Coffee Agreement. State support was dwindling rapidly as the Mexican Coffee Institute stopped its marketing regulation functions. The neo-liberal economic

---

[27] Currently the general director of the export company Comercializadora Más Café.
[28] In the middle of the export season and while successfully implementing the operational plan, the marketing team found out that newly hired employees were developing a parallel business plan for the Unión. The team went on strike in April 1999, to negotiate payment of wages through their newly established consultancy. Furthermore, following the success of operating the business plan for the commercialization of green coffee, they offered to extend their services to the other enterprises within the Unión, like the processing plant, transport company and roasting plant, for a lump sum per month. A decision regarding this offer was postponed and set on the agenda for the General Assembly in June 1999. Meanwhile the marketing team resumed their activities. The Assembly in June decided not to hire the consultancy and asked the marketing director and accountant to stay. After some initial talks, both declined and started to work in KAFFE with the other three ex-employees of the Unión.

politics of Salinas de Gortari and the Free Trade Agreement with the United States and Canada (1994) resulted in a general withdrawal of the government from market regulating functions. National politics were changing rapidly around 2000, when elections resulted in a president from the opposition for the first time in more than 70 years. This changed the character and need for political negotiations. The groups working with KAFFE did not start their joint venture under the same adverse political conditions experienced by the Unión de la Selva at the outset of its commercial operations. Not only had political conditions become less adverse, but the road network in Chiapas improved considerably after 1994, facilitating logistics and communication.

For all farmer organisations, with the retiring government in the public domain and traditional clientelistic relations fading fast, market partners were becoming more important as a source of financial autonomy for both organisation and their members. In other words, the conditions for external adaptation had changed.

10. Internal integration

As already mentioned, largely the same organisational culture can be ascribed to both the Unión de la Selva and to the farmer groups that started to work with KAFFE in 1999. They shared dynamics of social cohesion at community level and history as participants in social farmer movements. The two cases, Unión de la Selva and Comercializadora Más Café, will now be compared on the basis of the following question:

What elements from the strategy implemented by KAFFE influenced the organisational culture of the organisations part of the Comercializadora Más?

In relation to this question, Alblas describes five factors influencing organisational culture:
- awareness of the need to change;
- vision, commitment and inspiration from the top;
- broad organisational and structural support within the organisation; training and confirmation of desired skills; and
- the introduction of new symbols and rituals.

11. Unión de la Selva

Because of the financial problems and immediate intervention in operations by an external auditor, there definitely was an awareness of the need to change within the Unión de la Selva. Nevertheless, the immediate focus of these changes was on administrative systems to solve financial problems. There was less commitment to organisational changes as externally introduced budgetary discipline limited internal financial and political room for manoeuvre. This led to the board of directors being sandwiched (and internally divided) between the auditor and formally signed agreements on the one hand, and the passive resistance and evasion of agreements by other internal actors on the other. The imbalance of informal power relations led to an internal climate of insecurity in which the board of representatives of the Unión was reluctant to learn and become accountable.

Broad support for change from members was not automatic as they were ignorant of the complicated web of entities in which the Unión was operating. Never before had they decided upon internal control systems or demanded a formal definition of rights and obligations. When these were demanded by FLO[29], the Union initiated a round of talks with all associated communities to define and formalize rights and obligations for members. The majority of members supported the initiative, even if it meant assuming financial liabilities. Nevertheless, this process did not lead to formal agreements.

The employees lacked opportunities to connect regularly with members at the grassroots level to explain what they were doing vis à vis the implementation of new systems, and how[30]. The success of business management tools (an operational plan, a budget, quality control etc.) was ignored by the board or not communicated, and was largely unknown to the majority of the members. As a result, in and after the crisis, members kept going back to the people they trusted most to judge the veracity of information provided to them. In other words, no major changes occurred within the 'old' organisational culture.

Likewise, the departure of part of the staff and some of the communities originated from organisational (dis)function in relation to the external environment. Founding communities since then have reacted with feelings of betrayal to the departure of their staff and some members, particularly towards those with whom they most identified (Gonzalez 2001, Estrada 2006). It would seem that when those members and staff left the organisation, they were considered to

[29] Fairtrade Labelling Organizations
[30] With the exception of one group: Kulaktik was informed in detail about the export of a container of their coffee to a buyer in the United States who was setting up an exclusive coffee brand and specific relationship with the group through mutual parishes.

have breached the values of loyalty to (founding) members and affinity to the organisation[31].

12. Comercializadora Más Café

Having been hired as an independent consultancy under contract, when introducing new systems, KAFFE did not personally threaten political positions within the farmers' organisations. All the same, KAFFE was not an external element in the organisational culture of any of the groups, having a proven track record in coffee and with farmers' organisations in Chiapas. Actually, KAFFE can be considered as another kind of incorporated external agent. The trust instilled in KAFFE by the groups enabled it to initiate and set the agenda for coordinating commercial activities. On the other hand, not all groups equally understood KAFFE's new role as an 'internal broker'. There were occasional attempts to create mutual moral dependency by offering presents or creating goodwill beyond the practice of hospitality.

There was a clear common mission: the reduction of costs; the foundation of a joint export company; and, the conquest of new markets. The board of representatives and natural leaders were key elements in gradually changing internal attitudes and values. In the process of learning about markets and management systems, they gradually conveyed to members of their organisations where the priorities of market relations lay and that members themselves have a direct influence on the quality of their coffee. As they started to implement business plans, the newly introduced transparency and checks and balances limited the space for trade-offs. With regular visits to assemblies and members, KAFFE could countercheck the internal information flow and organisational functioning of board and staff, and facilitate understanding of procedures at the grassroots level.

Access to grassroots level within Comercializadora Más Café, as within La Selva, was more difficult when the organisation was already successful politically or commercially. In those cases, internal leaders had more to lose. Otherwise stated, the less the entrepreneurial experience in the organisation, the more eager its leaders were to

---

[31] A small faction from one of the founding communities give a rather distortional account of events, recurring to defamation of personal reputations and false accusations of theft and fraud, mainly directed at members from the founding communities who left the organization, and at the employees who formed KAFFE (see Estrada 2006).

learn 'the tricks of commercialization', seeing this as an opportunity to strengthen their position internally.

The support at grassroots level was based on groups wanting to see their own organisation exporting independently. In fact, each group exported under its own name during that first year. Though all groups understood the economic benefits of joint operations, some preferred autonomous operation despite the higher costs[32] incurred.

The joint training programme carried out by KAFFE, generated more interaction between groups and created trust amongst group representatives. In addition, training was fun and provided an occasion to socialize. Several farmers took the opportunity to specialize professionally within the organisation. They are now directors, cuppers and quality managers. This is a new generation of 'advisors' - externally and internally trained professionals - who are familiar with the politically sensitive environment of producer organisations and have an affinity with and authority among organised farmers.

Finally, Comercializadora Más Café has been a success for the participating organisations. Their new joint identity is parallel to and does not substitute the identity of each separate group with particular religious, ethnical and political characteristics.

13. Conclusions

Competitive participation in an international supply chain requires effective and continuous communication; swift decision making at critical moments in the negotiations with buyers or when the market fluctuates. Commercial activities demand sound financial administration, clear resource allocation and clear definition of the relationship between members and organisation.

Organisational culture may cause an organisation to function differently from its formal structure or as required by the market in terms of efficiency and financial transparency. It reflects the history of the producer organisation and its identity. In the case of Chiapas that history is characterized by indigenous traditions at the community level, by links of political patronage between the State and farmer

---

[32] The organization Productores Orgánicos de la Sierra Madre (POSI) started as one of the six groups working with KAFFE, but after the first year decided not to form part of the Comercializadora Más Café. The organization Tzijib'Babí was one of the founders of the Comercializadora, but sold its shares after three years (2003) to start operating on its own.

organisations, and by values of loyalty and trust between farmers and the people working with them.

When comparing the Unión de la Selva with Comercializadora Más Café in the implementation of market management systems, all groups in both organisations shared the lack of internal accountability and formal definition of rights and obligations by members. They shared strong principles of solidarity and consensual agreements at the grassroots level and have integrated external agents to facilitate negotiations with their external environment.

The first major difference that influenced changes in organisational culture, relates to when in the organisational life cycle the changes were introduced. The Unión de la Selva already had a successful history of combining political negotiations with commercial activities and a strong reputation in the international market, but was forced to restructure when its way of operating led to severe financial crisis. The changes created resistance as internal power relations were affected. Lack of sound information about current affairs led to speculation at the grassroots level, growing distrust and loss of personal reputations within the organisation. This led the organisational culture to become rigid and defensive.

By comparison, the groups working with KAFFE were just at the start of creating a new joint venture. They had a common mission: to gain market access and administrative skills by establishing joint financial-administrative systems and clear mechanisms of internal control. Through KAFFE, mutual trust was built among the organisations during operations and backed up in the formal agreements that formed the foundation of Comercializadora Más Café.

A second difference that influenced changes in organisational culture, relates to the fact that contrary to the case of Unión de la Selva, in the case of the groups associated with Comercializadora Más Café, KAFFE was able to establish a programme of producer support at all organisational levels. Internal leaders balanced the benefits of 'learning the tricks' with the inconvenience of having KAFFE intervening in internal affairs. Training included staff, natural leaders and board members.

The practice of farmers using business tools, in particular a business plan with a congruent organisational chart and delineated responsibilities, supported groups in finding alternate ways of functioning. In the process, a gradual change in organisational culture and perceived identity took place. Representation changed from being primarily a function of honour, privilege and positions assigned on the basis of seniority and authority, to a being more instrumental and related to entrepreneurial leadership, knowledge and skills.

Finally, with successful market integration, access to resources changes from being related to external state policies and successful participation in NGO networks, to being generated by members themselves. This gives members more autonomy vis à vis external parties and allows internally for negotiation as to voting rights and property rights over joint assets and investments.

References

Douglas, T., 1995. Over leven in de groep, psychologie van het groepslidmaatschap, Uitgeverij Intro, Baarn.

Estrada Saavedra, M.A., 2006. Entre Utopía y realidad: Historia de la Unión de Ejidos de la Selva, Liminar. Estudios Sociales y Humanísticos, Junio, año / Vol IV, número 001, Universidad de Ciencias y Artes de Chiapas, San Cristóbal de las Casas, México, pp 112 - 135.

Fox, J., 1994. The difficult Transition from Clientelism to Citizenship: Lessons from Mexico, Wordpolitics, Vol. 46, No.2, pages 151-184.

González Cabanas, A., 2001. Evaluation of the current and potential poverty alleviation benefits of participation in the Fair Trade market: The case of Unión La Selva, Chiapas, Mexico, fair trade research group, Colorado State University y Desarrollo Alternativo, A.C.

Harvey, N., 1998. Chiapas Rebellion: the struggle for land and democracy, Duke University Press, Durham and London.

Honda, H., 1997. Political Process Approach: A Brief Review of a New Perspective on Social Movement Research. The Hokkaido Law Review (Bulletin of Universities and Institutes), 48:3, p. 226-250.

Legorreta Díaz, Ma. Del C., 1998. Religión, política y guerrilla en Las Cañadas de la Selva Lacandona, Cal y Arena.

Leyva Solano, X., 2001. Regional, Communal and Organisational Transformations in Las Cañadas, Latin American Perspectives 2001, vol. 28, pages 20-44.

Mattiace, Shannan L., 2001. Regional Renegotiations of Space: Tojolabal Ethnic Identity in Las Margaritas, Chiapas, Latin American Perspectives, 28, 73

McNamara, C., 2005. Field guide to Consulting and Organisational Development with Nonprofits: A Collaborative and Systems Approach to Performance, Change and Learning, Authenticity Consulting, LLC, Minneapolis, MN.

Osten, E. van, 2006. Authoritarian durability and democratic transition in Mexico, University of Virginia, Woodrow Wilson Department of Politics, paper prepared for 2nd Annual Conference, March 10, 2006.

Parker, M., 2000. The sociology of organisations and the organisation of sociology: some reflections on the making of a division of labour, The Editorial Board of The Sociological Review 2000.

Schein, E.H., 1992. Organisational Culture & Leadership, 2d. Ed. San Francisco, CA. Jossey-Bass.

Peer-to-peer farmer support for economic development

Kees Blokland and Christian Gouët

> *Expert farmer-to-farmer advice is one of Agriterra's organisational support strategies. This chapter describes several cases in which one advantage of peer-to-peer missions is illustrated: that of generating tangible benefits and services for members.*

1. Introduction

Agriterra's direct goal in development cooperation is to strengthen farmers' organizations (Blokland and Gouët, 2007). This chapter will focus on a specific form of advisory service that was developed as a cornerstone for development cooperation practice: peer-to-peer advisory services. We explore how these services relate to the aim of Agriterra's development cooperation: i.e. to strengthen farmers' organizations for development impact. This chapter provides new input derived from 10 years of experience mobilizing experts from private enterprise, cooperatives and farmers' and rural women's organizations. In building up Agriterra's peer-to-peer advisory service, the so-called 'AgriPool', was the first peer-to-peer support method to be institutionalized. Development impact is widely understood to stem from the implicit drive for innovation, which we relate to the diffusion of technological innovations, especially market-led, wherein farmers buy solutions from companies seeking to meet the specific needs of their clientele.

Innovation theory emphasizes that more radical forms of innovation require change at multiple levels and locations simultaneously (Smits, 2002). That producer organisations are organised at different levels is a potential advantage from this perspective. At the local level they can, for example, organise necessary inputs, facilitate collective management tasks or safeguard transport and storage. At the national and international levels they can create space for innovation by negotiating trade agreements and conditions with governments or industries. At intermediate levels they can play a role in establishing facilities for processing, grading, certification and service provision. Services can be provided to farmers either as a product of lobby activities attracting the attention of service providers or through the formation of co-operatives developing the service as a farmer-led enterprise or through partnerships with third parties like private companies. In playing their many roles, farmers' organisations can build on best practices from peers around the world. Contacts between peers establish valuable additional relationships for farmers' organisations, along with the kind of relations normally within their scope of interaction: relationships with their members, markets and external stakeholders, like governments. We argued that peer-to-peer contact is an important piece of the development puzzle that allows information exchange regarding technology, markets and other experiences. Hence, the facilitation of peer-to-peer contact is central to the development practice of agri-agencies[33].

2. Peer-to-peer advisory services

AgriPool is a business unit at Agriterra that is run as a non-profit employment service. The Agriterra liaison officers make up the front office and work to match demand and supply. The advisory services of AgriPool are exclusively rendered by (former and active) directors, employees and regular members of farmer unions, cooperatives and associations of rural women. AgriPool comprises all activities, persons

---

[33] For more information on agri-agencies, please refer to box 1 on page 240.

and systems comprising Agriterra that make professional advisory services in developing countries possible. AgriPool recruits advisors from rural people's organisations, both Dutch advisors participating in Agriterra, and advisors from rural people's organisations around the world. AgriPool deals with the whole mission cycle starting with the initial request from a rural people's organisation in the south until the final report, feedback via magazines of the sending organisation and via the press, and the evaluation.

AgriPool is based on five leading principles:

1. The aim is to render professional advisory services.
2. Peer-to-peer advice brings together people with similar positions in comparable organizations.
3. AgriPool experts are preferably deployed with or in preparation for long-term cooperation between one or more producer organizations.
4. Experts are required to divulge the results of each mission to broader audiences, beyond those directly involved.
5. Finally, AgriPool services are voluntary contributions, i.e. without payment.

Every step of the mission must meet the standards of a professional consultancy firm: expert selection based on specialization and skills, formulation of the Terms of Reference, reporting requirements, and job preparation. The experts chosen are directors, staff or regular members of farmers' organizations and cooperatives who possess concrete expertise in the area in which advice is required.

Peer-to-peer advice brings colleagues together in a way that reflects the principles of popular education[34] and can be better to understood as intervision[35], rather than formal one-way advice. Deploying experts in long-term cooperation guarantees that knowledge is accumulated within the farming community worldwide. Missions generally involve a sending and a receiving organization; both are rural people's organizations.

---

[34] Popular education evolved as a means to overcome exploitation and social alienation. Paolo Freire (1972) talked about ordinary working people wanting to learn skills relevant to their harsh and oppressive lives. Social change begins with individuals reflecting on their values, their concern for a more equitable society, and their willingness to support others in the community. See the work of Peter Reardon about popular education and learning for social change, on line on: http://adulted.about.com/cs/learningtheory/a/pop_education.htm

[35] Intervision is a type of capacity building in which employees call upon colleagues in identical positions from other companies to discuss personal, function-related or professional issues arising in their daily work.

The experts, as well as the sending and receiving organizations publish results in their magazines and in regional papers. This generates over one hundred press clippings per year.

The voluntary character of the missions does not necessarily mean that the missions are low cost: the total cost tends to be below that of typical expatriate expert missions, but higher than the deployment of local experts. Some experts require a vacancy payment in order to hire someone to replace them on the farm or to allow their organization the opportunity to find a replacement during their absence. We calculate a standardized fee of € 250 per day for this type of service.

The average cost per expert on mission is € 6.500, excluding the activities of Agriterra liaison officers for preparation and debriefing. As pocket money, experts receive only 14% of the official UN Daily Subsistence Allowance. The rest is paid to the receiving organization to cover the logistics. The sending organization or expert can claim compensation if they need to hire a replacement on the farm or within the organization. Increasingly, this compensation is donated to Agriterra and then used for additional project funding.

Advisor recruitment is done in two ways: 1) open requests in the magazines of Dutch farmers' and rural women's organizations; and 2) applications for concrete vacancies for missions that are advertised on Agriterra's website. Open applications are screened and evaluated based on the candidate's technical and organizational abilities, work experience in an international setting and the possibility for feed back from the sending organization. Language proficiency is tested by a specialized bureau.

Applications are directly submitted to the liaison officer who evaluates the qualifications of the applicant relative to requirements for the mission. When needed the liaison officers undertake an active search for particular expertise within the Dutch rural people's organizations.

A considerable number of the approximately 180 experts in the pool have never actually been sent on mission. This is largely due to the relatively stable number of missions undertaken annually - approximately 60 per year until 2005. Another factor is that the principles of AgriPool prioritize experts who are involved in collaborative arrangements between Dutch and foreign organizations. This leads to many of the same experts being deployed on different (often follow-up) missions. The large number of experts found in the database compared to the demand for particular expertise is an asset from a recruitment perspective, but has led to some frustration amongst experts in the pool. This may develop into a problem as it restricts the possibilities for the rural people's organisations in the Netherlands to transform commitments with colleagues in the developing world into real action. An increase in the number of missions (as is foreseen for the years ahead) will partly resolve this problem. In addition, regular re-evaluation of the expertise found within the database compared to expected demand is necessary in order to prevent further frustration.

Missions over the period 2001-2003 fell short of expectations. Only 70% of the projected 260 missions were realized. In that period, particularly in 2002, a considerable number (30%) of experts from developing countries were deployed. Some were more closely linked to daily advisory practice and worked as Chief Technical Advisors, later referred to as 'PO-advisors'. In 2006, the number of advisory days for farmers' organizations in the developing world met expectations thanks to the work of these PO-advisors. In that year, while only 89 of the 160 missions planned were actually realized, almost 96% of the intended 2 300 advisory days were delivered.

In the early days of Agriterra, missions were exploratory in character aiming to establish relationships and build a foundation for international collaboration. Many collaborative arrangements emerged from this period. In 2005 almost 45% of the farmers´ organizations in Agriterra's programme collaborated with a Dutch organization. Nowadays, most missions are clearly advisory in nature. Most experts (55%) are from farmers' organizations, with a smaller proportion being from cooperatives.

The administrative handling of missions was partly an in-house undertaking and partly arranged through an external bureau that dealt with providing pre-mission information to experts, report editing, and updating information about each mission on the website (see www.agro-info.net). This new website is an updated version of the peasantworldwide.net website and is still under development as a management tool for AgriCord. According to early objectives of Agriterra, the website is also intended to act as an exchange mechanism amongst farmers' organizations providing easy reference to previous experiences and lesson learned. News about missions that is published Agriterra's website is prepared for the most part (90%) by 'news service' volunteers. These communications, in addition to contributions submitted by the experts to their own organization's periodical or local newspaper, are generating enthusiasm and commitment amongst a far broader audience than those directly involved.

Producer organisations and market chains

3. A first evaluation of peer-to-peer advise

Evaluation of the AgriPool in 2005 concluded that more attention should be given to competence management, new interview and screening techniques and evaluation from a users' perspective. External use of AgriPool experts is still weak. Most agencies are not familiar with the use of this type of expert, and they normally fall back on more traditional development experts, both expatriate and local. However, there are clear advantages to the AgriPool approach compared to local or expatriate experts.

The prototype expert was Toon Bierings, a former high ranking official at Rabobank in the Netherlands, who began in 1992 to work with the organisation that preceded Agriterra. He travelled on multiple occasions to Nicaragua to help establish a rural bank, *Banco del Campo*, of which he became a Director, representing foreign investors (Vermeer, 2005). On all missions Bierings requested that he be accompanied by an experienced development expert and by local experts, either from within the organisation or sometimes externally recruited. He considered the local receiving organization to be fully responsible for the results of the mission, with his own contribution being to provide additional input to support their efforts. Work planned was duly described in the terms of reference. Every mission concluded with a report that contained recommendations for follow-up, with Bierings being the key player in follow-up actions targeted towards foreign parties. He organized a broad spectrum of agencies and banks that in a 'joint venture' supported Banco del Campo's planning and investment phase and in the early years of its existence. For personal support in these endeavours, Bierings depended on Agriterra.

Bierings repeated his successful approach in Costa Rica, Peru and to some extent in Uganda, India and Madagascar. He was an example for the local Rabobank manager, Jo Opsteegh, and many others from LTO[36] organizations and rural women's organizations.

From these successful missions, several success factors emerged:
* personality of the expert;
* professionalism;
* deployment in missions that match the interest and expertise of the expert;

[36] LTO is the Dutch Organisation for Agriculture and Horticulture (Land- en Tuinbouw Organisatie Nederland) in the Netherlands. It is an umbrella organisation for five regional and sixteen sectoral organisations in agriculture and horticulture.

- using local expertise on the issue and on the socio-political environment;
- embedding the mission in local action that supporting the objectives of that action;
- clear description of tasks and objectives;
- hands-on follow-up by the expert;
- support from the Agency in follow-up.

Reviewing the 180 Dutch AgriPool experts in light of these success factors leads us to conclude that the strengths of the Agripool experts are: personality, professionalism, matching interest and expertise to the mission, clear terms of reference, and the support they receive from Agriterra. To date the weaker points have been: insufficient collaboration with local expertise; a failure to embed the mission in the local action of the receiving producer organization; and poor follow-up by the expert after the mission, including the composure of clear, succinct and readable reports.

In cases where missions were not embedded in local action and in the absence of a local group within the farmers' organization participating in the mission, the link to clearly identified local expertise was lacking and follow-up suffered. The expert then identifies much less with the local organization and is less dedicated in terms of raising support, post-mission. A review in 2005 showed that of 110 experts on mission, 48 had been on more than one mission and 21 had been on more than two. A repeated mission by the same expert is to some extent a sign of successful follow-up. By the end of 2006, Bierings' approach had been taken up by approximately 30 experienced experts.

4. Cases

We will now share a few stories connected to this group of 30 experts to illustrate both the advantages and the effectiveness of Bierings' approach. For Agriterra's fifth anniversary celebration, several cases were filmed and presented: 'Pierik-Interpolis-Mutual Insurance' (Philippines-Nepal); 'Opsteegh- programme council-ranchers' (Nicaragua); and, 'Van Vossen-LTO-potato chain' (DR Congo)[37]. The messages that respectively emerge from these stories are: 'we have excellent organizational solutions and methods developed that can be of great use for developing countries.

---

[37] Three filmed accounts produced by Lokaal-Mondiaal for Agriterra, Arnhem, 2003

We are ready to put our expertise at your service'; 'we can offer a complete set of support covering general advocacy for farmer interests, women's issues and support to agricultural, cooperative and business development including banking'; and 'we are ready to make longer term investments working along with you over many years, because we know that major changes do not happen overnight.'

In this article we will describe an additional three cases:
- 'Van Bohemen-ZLTO[38]-Starting a farmers' organization from scratch' (Thailand)
- 'Van Rossum-WLTO[39]-downward accountability and outward representation' (India)
- 'Schutte-NLTO[40]-cotton advocacy and pro-poor activities' (Benin)

4.1 Thailand

The involvement of farmers in the south of the Netherlands with those in Thailand goes back to the late 1970s and 1980s. This was a period when development agencies were criticising the import of cassava as a feeder crop for Netherlands cattle. The imbalanced growth of the Dutch national herd and the corresponding manure and environmental problems evident resulted in the calculation of a 'global footprint' for cows in the Netherlands: an estimated seven hectares outside the Netherlands were needed to feed cows grazing on one hectare within the Netherlands.

Initial contacts in Thailand centred around a Dutch expatriate and support from an NGO, namely Cordaid. In 1998, ZLTO approached Agriterra in order to coordinate its activities in Thailand. From the situation observed by the Agriterra liaison officer we learned that the institute supported by ZLTO was not a farmers' association, but in fact an NGO connected to the Ministry of Agriculture. To build and support an independent farmers' organization, which was the intention of both ZLTO and some of the Thai partners involved, two steps were required: breaking away from the ministry, and bringing the NGO under democratic farmer governance.

Kees van Bohemen visited the NGO approximately a dozen times over the period from 1998 until 2006 and was involved in this transition. He advised the NGO by sharing experiences related to the development

---

[38] ZLTO is the South regional branch of LTO
[39] WLTO is the West regional branch of LTO
[40] NLTO id the North branch of LTO

of a farmers' association in the southern Netherlands. On return visits, ZLTO branches illustrated their policy preparation, negotiations and advocacy at all organizational levels, and included their guests as observers in meetings and special workshops. Strategically strengthening farmers' organization in ten districts, FAD (the organisation's name at the outset) managed to organize approximately 20 000 farmers and rural women, mostly through clustering existing associations. Through well directed training on leadership and the election of directors at all levels, a solid organization was built. Agriterra assisted in the building up of the national secretariat with function descriptions, procedures and communication system.

The relationship with Thailand is well-embedded in ZLTO. Its president and several directors as well as employees have visited the country. Missions and return visits are described in some detail in *(De Nieuwe) Oogst*, the magazine of LTO. Member contributions are collected locally and as of 2005 local branches began contributing to the upper levels of FAD. All offices both central and district have computerized member registrations. The building up on district level received government support. Advocacy regarding the increasing indebtedness of farmers resulted in a major debt reduction scheme and a strong focus on savings and credit institutions. In fruitful collaboration with the Credit Union league of Thailand (CULT) and with technical advice from ACCU, the Thai-based Asian Association of Credit Unions - SorKorPor - as it was re-named, became a member of the newly established Asian Farmer Alliance. In the years ahead, SorKorPor will be a frontrunner in the effort to focus activities and related development cooperation to the lower levels of the organization, thereby strengthening the grass roots level.

The central lesson from this case is that Agriterra liaison officers' development cooperation expertise and the specific organizational expertise of ZLTO make a powerful combination.

4.2 India

In 2001 Toon Bierings and Ria van Rossum visited India on an exploratory mission in order to evaluate the potential of the Andhra Pradesh Federation of Farmers' organizations. Van Rossum is a woman farmer and member of the regional LTO organization in the West of the Netherlands (WLTO). At that time she held the position of Director and was President of the LTO branch of the commission for development cooperation.

Later she became President of the advisory council of Agriterra. In her five years in office she actively recruited AgriPool personnel for many assignments. She was also the instigator of collaboration with the WLTO in Central Africa that provided support to organizations in Rwanda, DR Congo and Kenya. In that sense, this case in India was not Rossum's main contribution, however, it does illustrate the impact that these types of experts can make.

Van Rossum's mission concluded that FFAP had potential, particularly due to its influential and dedicated president. At the same time it lacked close ties to its members. In 2002 she invited President Chengal Reddy for a return visit for the occasion of World Food Day. The relationship with WLTO helped FFAP to define itself as a farmers' organization and break away from the idea of building an NGO-type of institution for providing farmer support.

From that point on, FFAP started to work with Agriterra to consolidate the federation through building linkages with 200 existing farmers' associations. Van Rossum assisted FFAP in formulating the support project and recruited the head of the training unit at WLTO to give leadership training to its directors. A few years later, FFAP had established 247 local associations, and trained over 1200 women farm leaders for income generating activities. More important for the future of the federation were the external linkages institutionalized with parliamentarians (Agricultural Forum of Congressman) and with industry (Indian Farmers and Industry Alliance). The influence of FFAP spread to credit, access to water and, protection of the home market, amongst other issues.

Within a few years, guided by the example of farmers' organizations internationally, FFAP had built a respected federation with links to many farmers' organizations in other states. The state plan for agriculture cannot be approved without FFAP having commented in the press. The central issues of the ever expanding FFAP were: advocacy, business development, and effective service provision to members especially on agricultural technology (FFAP is a fervent advocate for biotechnology and agricultural innovation).

Developments at this point were no longer triggered by Dutch input. Instead Agriterra started to contract the expertise of Anil Epur - the person who was responsible for FFAP connections with industry - as PO-advisor. Of Epur's many contributions, his involvement in introducing colleague organizations in new ICT solutions. Andhra Pradesh is Asia's Silicon Valley. Some FFAP mango producing associations enjoyed new success when their productive improvements, networking and advocacy enticed Coca Cola to contract them as preferred suppliers for 'Frooti', the number one mango drink in India. Another success was the establishment of the Confederation of Indian Farmers' Associations, an initiative actively promoted by FFAP with the support of Agriterra and Anil Epur. These formal farmer alliances bring the total membership in farmers' organisations to 120 million Indian farmers. Read this figure carefully and reflect on it!

This case illustrates that the Dutch example provided farmers in India with a mirror to their own organizational future. Chengal Reddy looked into that mirror and knew what to do.

4.3 Benin

Ecooperation, the Dutch foundation that managed the partnership agreements for sustainable development signed with Costa Rica, Bhutan and Benin, encouraged Agriterra in the mid-1980s to work with the farmers of Benin. The first action was an inventory, consulting a dozen Dutch NGOs and institutes with a track record in Benin in order to assess the landscape of existing rural people's organization. SNV[41] provided information on grass-roots level associations and several women's groups. Apart from SNV, the people consulted indicated that farmers were unorganized. This diagnosis is an example of NGO blindness prevalent in development cooperation at that time, for genuine organization of the people and biased attention towards local NGOs. A short visit and diagnosis conducted by Agriterra laster revealed that there was in fact a farmers' federation with national aspirations (FUPRO) and a Chambre d'Agriculture in the country (Agriterra, 1998; FUPRO *et al.*, 2000). The same study illustrated that this farmers' organization remained weak in the poorer zones south of the cotton belt.

With Agriterra support FUPRO started to work in these poorer regions. This was important because the initiators of FUPRO are cotton farmers. As cotton is an export product that is also produced on larger farms, FUPRO had been branded a rich farmers' federation. In these region FUPRO started to improve cassava, palm oil and pineapple[42]

---

[41] SNV is a Netherlands-based, international development organisation that provides advisory services to nearly 1800 local organisations in over 30 developing countries (www.snv.nl).

[42] AFDI, the French agri-agency, illustrated the results in the latter case during the 'Farmers Fighting Poverty' seminar in Arnhem, 2006.

chains also. The NLTO deputy president Willy Schutte and policy officer Douwe Hollenga initiated cooperation with this Northern LTO branch with FUPRO and brought their attention to the cotton sector. The possibilities for organic cotton were commonly explored. Directors, staff and members of FUPRO visited the Netherlands on numerous occasions for exchanges with LTO and French farmers or to present vegetables at Rotterdam's AGF fair. NLTO staff commented on FUPRO's plans to start up a demonstration farm for sustainable agricultural production. In 2003 NLTO directors were involved in reorganisation discussions at FUPRO that ultimately gave the cotton growers branch status and resulted in their incorporation to the FUPRO board.

In 2002 Klaas Jan Osinga visited FUPRO for the first time to participate in a conference on organic cotton. This niche product was presented as a means to overcome the low price cotton crisis. Low prices were attributed to subsidized exports from the United States, the second biggest cotton exporter in the world. Indebtedness of cotton producers was the result. Yet, this study did not provide compelling evidence that organic cotton would be a solution for the cotton sector in Benin. The monopolistic position of the state ginnery was a more urgent issue to address.

This conclusion is in contrast to the conclusions of Sinzogan *et al.* (2006). According to their findings, FUPRO is an obstacle for innovation and 'development'. They say that FUPRO is part of the problem facing Beninois farmers rather than part of the solution. They state that opposition to organic cotton is due to the fact that FUPRO has stakes in inputs deliveries for cotton. 'A common perception of the N'dali farmers interviewed is that: [in French] the secretaries [GV-Fupro] are those that ruin us, they are the only ones who pull profit out of all the cotton production.'...(Sinzogan *et al.*, 2006). The authors judgement appears to stem from disappointment that FUPRO did not make the authors' preferred technological choice. However, from a development perspective it is more important that farmers' organizations promote collective thinking about technology and trigger their members to make choices that suit their farm and family (Blokland, 1992) according to market possibilities and constraints.

FUPRO cotton growers need to confront market realities with efficient and sustainable technical approaches. FUPRO is not convinced that it should promote organic cotton. The availability of organic sources of fertilization is a sensitive issue. In addition, it is difficult to forecast future market opportunities. Benin produces about 0.5% of total global cotton production (USDA, 2007), and organic cotton is about 0.05%. It is difficult to find convincing statistics related to the organic cotton market, however based on the information we have, shifting Benin's cotton to organic would result in a ten-fold increase in the world supply of organic cotton. Smallholders in Benin therefore need information and contacts to assist them in judging the dynamics of the cotton market and their future choices.

During the IFAP[43] World farmers Conference in 2003 in Washington, Osinga introduced FUPRO directors to the National Cotton Council of the United States. On that occasion two insights emerged for FUPRO directors: the first was that in order to regain market share, carefully designed marketing was needed to create an unique selling point for West-African cotton; and, the second was that developments in major cotton-producing countries such as China and India were far more important in terms of dictating the opportunities for West-African cotton in the world market (including the organic cotton market) than United States' subsidized cotton.

The case of FUPRO illustrates the benefits of contacts made through international cooperation, especially when they go beyond one country, as in this case where links were explicitly made to Dutch and French efforts and placed in the IFAP framework.

5. Advantages and impact

The six cases referred to in this article, all highlight specific advantages of peer-to-peer advice. The AgriPool experts offer valuable and proven tools and organizational solutions and are able to tap from a wide array of experiences and expertise in virtually every field. Dutch history offers a vivid example of the development impact of farmer organization and farmer-led economic enterprises. Experts from the agrarian sector bring that history with them. Once on the ground they build commitment and companionship in long-term arrangements that link development cooperation expertise. Farmer-led development cooperation combines the socio-political and economic force of farmers and cooperative enterprises worldwide with the development cooperation expertise and influence of the agri-agencies. In that way, it opens doors and provides high level contacts.

---

[43] IFAP is the International Federation of Agricultural Producers

However, peer-to-peer advice is not a panacea. Due to cultural, technical and scale differences, on some occasions examples provided by European farmer experts are too abstract and distant from the reality in developing countries. Van Hoof and Destrait (2006) recommend for instance the use of emerging regional farmer platforms, like the East African Farmers' Federation. Stronger farmers' organizations can serve as examples to the weaker ones. Likewise, they recommend increasing the possibility of learning via the internet about best practices and the benefits of South-South peer advice. In its strategic plan, IFAP stresses the need to identify and mobilize peer-to-peer advice worldwide, thus not restricting advice to only north-south exchanges. Regardless, AgriPool experts are just one of the possible knowledge inputs available to rural people's organisations in developing countries. This chapter illustrates their value and some positive results to date.

Farmers Fighting Poverty program projections are that the number of missions will increase from 150 in 2007 to 461 in 2010. This follows the trend of a general increase in activities intended within the programme. The programme will be active in projects initiated by farmers' organizations directly involving 2.7 million farmers as active participants. These farmers belong to associations and federations with a membership of almost 25 million. This represents about 10% of the total organized peasantry worldwide: an estimated 250 million farmers out of a rural economically active population of 1.2 billion. This 250 million is linked to the development efforts of national, regional and worldwide bodies to which these associations belong.

It could be argued that involving only 1% of the organized farming sector in development cooperation activities or bringing the expertise of 34 full time equivalents cannot seriously be thought to spur economic development, bring democracy or reduce poverty. However, it must be remembered that this 1% of participants represents entrepreneurs carefully selected from among the total membership of the organizations involved. Ryan and Gross's (1943) study on the diffusion of agricultural technology found farmers involved in exchanges to be innovators and early adopters. The description of early adaptors in particular refers to the type of farmers, members of open associations, defined by Paxton (2002) as: associations that have members who maintain multi-stranded relationships and form part of different institutions, fostering social capital and development processes. The early adopters in Ryan's and Gross's study were typically respected members of the rural community and often played dual roles as both farmers and role models in banking, real estate, government, educational or religious institutions in the region. This group consisted of the most successful farmers, who were also considered respectful leaders in their communities. Combining theories on entrepreneurship, human and social capital with the economic theory of technological diffusion (Benhabib and Spiegel, 1994; 2003), may allow for an understanding of the power of development dynamics within an organized frameworks, such as farmers' organizations.

Ryan and Gross's observations on innovation diffusion have been used as a framework to promote innovation in rural environments, focusing for decades on modernizing farmers. The Wageningen School of Communication and Innovation Studies (formerly the school of 'Extension' comprised most notably of Anne van den Ban, Niels Röling, Paul Engel and Cees Leeuwis), formulated, in contrast to the Ryan and Gross's theory of individuals adopting technology, the concept of multi-stakeholder settings for innovation, i.e. the social organization of innovation. They explicitly include farmers' organizations in those settings (Röling, 1992; Engel, 1995; Leeuwis, 2004). However, despite the fact that diffusion theory leads to a much criticized strategy for extension-led innovation, it certainly describes the underlying process taking place: the natural process of the diffusion of ideas.

So, incorporating and elaborating on Wageningen criticism, diffusion of inventions in a non-induced, unplanned setting - i.e. through the market - might profit from the existence of farmers' organizations, perhaps even more than extension-led innovation. That is our kernel of theoretical insight. This is particularly interesting in modern times where companies offer solutions tailored to the needs and desires of farmers at the bottom of the pyramid (Prahalad, 2006)[44]. National farmers' organizations offer a channel for communication and information due to the many linkages between members participating in associations, commissions, delegations and meetings.

To date we have harvested many stories on this subject, like the AgriPool expert Bert Sandee who introduced a rack to conserve onions in Niger. On a follow-up mission he found farmers using this rack 1500 kilometres from the site where he had introduced it. The farmers explicitly referred to the fact that they had learnt about the rack from

---

[44] This makes reference to *The Fortune at the Bottom of the Pyramid* by C.K. Prahalad (2006). The phrase "bottom of the pyramid" is used in particular in the development of new models of doing business that deliberately target the market of the poor living on less than $ 2 a day, typically using new technology.

their national federation that had originally learned about it from a foreign expert.

6. Closing remarks

Leeuwis (2004, p. 219) building on Scarborough *et al.* (1997) points out that farmer-to-farmer communication for innovation can be seen as an optimal way 'to use the available knowledge, experience and skills from a farmers community'. Main advantages are derived from the fact that amongst themselves, farmers speak the same language, literally and culturally and are faced with similar problems and constraints. Leeuwis mentions that this strategy is also valid in other contexts, in which communication for innovations is done among members of a 'group', commonly referred to as '*peer* education'.

We have presented several cases to illustrate the advantages of peer-to-peer missions. They work to build or strengthen the organization and bring tangible benefits and services to members, sometimes far beyond the scope of the missions and related projects. The missions (as part of the total package of activities of both the farmer organization and of the agri-agencies) have an undeniably positive impact on development, (for examples see Van Hoof and Destrait, 2006 and www.agro-info.net).

Agripool illustrates how the effectiveness of a peer-to-peer strategy for supporting farmers organisations' on their path to economic development. Advantages of farmer-to-farmer communication that have been analyzed for decades and from different perspectives (Ryan and Gross, 1943; Freire, 1972; Rolling 1992; Engel 1995; Scarborough *et al.*, 1997; Leeuwis, 2004) increase exponentially when farmers organize. To see the full potential, one has to add what Paxton (2002) demonstrated regarding the interrelation between people's participation in open organisations, and democracy and development. In the end, the experience of Agripool offers an exiting opportunity to observe the peer-to-peer support path to economic development. This demands further examination as a route for the promotion of rural development.

As we argued, much of the causal relation between organization and development remains undisclosed so far. With a growing body of experiences and their careful analysis, including theoretical research, Agriterra hopes to solidify the scientific basis for its work and eventually be able to know which organization are more likely be successfully in what kind of activities and in what circumstances; and which impact on development, democracy and the reduction of poverty.

References

Agriterra, 1998. Plattelandsledenorgansiaties in Benin. Agriterra, Arnhem.

Benhabib, J. and M. Spiegel, 1994. The role of human capital in economic development: Evidence from aggregate cross-country data, Journal of Monetary Economics, 34, 2, 143-73.

Benhabib, J. and M. Spiegel, 2003. Human capital and technology diffusion, Working Papers in Applied Economic Theory, Federal Reserve Board of San Francisco, 2003-02.

Blokland, K., 1992. Participación Campesina en el Desarrollo Económico. La Unión Nacional de Agricultores y Ganaderos durante la revolución Sandinista. Ph.D. Thesis. PFS, Doetinchem.

Blokland, K. and C. Gouët, 2007. Farmers' organisation route to economic development, Agri-ProFocus, this publication.

Engel, P., 1995. The Social Organization of Innovation: A Focus on Stakeholder Interaction. Koninklijk Instituut Voor De Tropen, Amsterdam (September 1997). 240 pp.

Freire, P., 1972. Pedagogy of the Oppressed. Herder and Herder. New York.

FUPRO / Agriterra / CIEPAC / AJF / CBDD / Ecooperation, 2000. Le mouvement paysan au Benin. FUPRO, Cotonou, Benin. (CIEPAC: Centre International pour l'Education Permanente et l'Aménagement Concerté. AJF: Agrarische Jongeren Flevoland. CBDD: Centre Béninois pour le Développement Durable). On line in: http://www.agro-info.net/ ?menu = documents&view = document&document_id = 49267.

Leeuwis, C., 2004. (with contributions by A. Van den Ban). Communication for rural innovation. Rethinking agricultural extension. Blackwell Science, Oxford.

Paxton, P., 2002. Social Capital & Democracy: An Interdependent Relationship. American Sociology Review 67, April, 254-277.

Prahalad, C., 2006. The Fortune at the Bottom of the Pyramid, eradicating poverty through profits. Wharton School Publishing, University of Pennsylvania.

Scarborough, V., S. Killough, D.A. Johnson and J. Farrington (Eds), 1997. Farmer-led extension. Concepts and practices. Intermediate Technology Publications, London.

Sinzogan, A.A., J. Jiggins, S. Vodouhé, D.K. Kossou, E. Totin and A. van Huis, 2006. An analysis of the organisational linkages in the cotton industry in Benin. In press in: International Journal of Agricultural Sustainability, accepted in April 2006. Also published as a chapter in: Sinzogan, A.A. Facilitating learning toward sustainable cotton pestr management in Benin. The interactive design of research for development. Doctoral Thesis, Wageningen University 2006 - The Netherlands.

Röling, N.G., 1992. The emergence of knowledge systems thinking: A changing perception of relationships among innovation, knowledge process and configuration. Knowledge and Policy: The international Journal of Knowledge Transfer and Utilization, 5, 42-64.

Ryan, B. and N. Gross, 1943. The diffusion of hybrid seed corn in two Iowa communities. Rural Sociology. 1943;8:15-24.

Smits, R., 2002. Innovation studies in the 21st century. Questions from a users perspective. In: Technological Forecasting and Social Change 69, 9, 861-883.

USDA, 2007. Cotton Supply and Distribution MY 2006/07, Table 05. United States Department of Agricultura. Foreign Agricultural Service. Visited 09 May 2007 in: http://www.fas.usda.gov/psdonline/psdgetreport.aspx?hidReportRetrievalName = BVS&hidReportRetrievalID = 852&hidReport RetrievalTemplateID = 3.

Van Hoof, F. and F. Destrait, 2006. Rural producer organizations for pro-poor sustainable development. Contribution to the World Bank workshop in Paris, 30-31 October 2006. Agriterra/SOS Faim, Arnhem/Bruxelles.

Vermeer, R., 2005. Farmers want access to banks. Agriterra's experiences in Nicaragua, Peru and Costa Rica. Agriterra, Arnhem.

Section B

Value chain development with producer organisations

How can cooperatives meet the challenges of agrifood supply chains?

Jos Bijman

> *This chapter discusses the functions of cooperatives in general, and the new challenges they face when partnering in (international) supply chains. After describing both traditional and modern functions of a cooperative, the author discusses a number of challenges that are particularly related to their (new) role in the supply chain, focussing on financing, corporate governance, member commitment and member heterogeneity.*

1. Introduction

Agricultural cooperatives are a common form of organization for marketing farm products, both in developed and developing countries. They were established in order to support the economic well-being of (small) farmers. Cooperatives do this by strengthening the bargaining position of their members, thus creating a countervailing power vis à vis their customers such as processors, traders and retailers. In addition, agricultural cooperatives organize the coordination required between producers, and other chain actors such as traders and processors. Small producers often lack access to market and technical information, as well as sufficient financial resources to improve production quantity and quality. Cooperatives can address both of these challenges, thus allowing producers the opportunity to improve their economic and social situations.

Farmers all over the world are facing changing market conditions. Not only are agricultural policies shifting towards increased market liberalization and targeting new priorities (such as environmental sustainability and food safety), consumer demand for food products is also increasingly demanding in terms of quality, variety and safety. In addition, the structure of food markets is changing with the rise and consolidation of supermarkets, which affects both production and marketing conditions (Reardon *et al.*, 2003; Shepherd, 2005; Henson and Reardon, 2005). Supermarkets generally have higher quality requirements; prefer to deal with a limited number of suppliers; and demand homogeneous products. Supermarkets also tend to rely on private grades and standards, which often replace or are additional to public grades and standards. Finally, supermarkets - as well as processors and wholesalers - have become more concentrated, which adversely affects the bargaining power of small producers.

As rural producers become more embedded in national and international agrifood supply chains, and as delivery conditions in these chains are mainly determined by large private actors, one of the challenges is to strengthen their economic position in these chains. This refers not just about countervailing power, but also about being able to comply with quality and delivery requirements from domestic and foreign customers. One of the main questions is how (small) producers can become and remain suppliers in these demanding supply chains. Joining forces in a cooperative can be a significant part of the answer.

Analysis in this chapter is primarily based on the experiences of cooperatives in developed countries. However, I am convinced that these issues are also relevant for cooperatives and other POs in developing countries. Most internal challenges are valid for member-owned firms all over the world. But more importantly, cooperatives in developing countries face similar challenges to cooperatives in the developed world, as they become partners in modern supply chains.

This chapter discusses the functions of a cooperative in general, and the new challenges that cooperatives face when partnering in (international) supply chains. It starts with a description of the characteristics of a cooperative. Being both an association and an enterprise makes a cooperative a special type of organization, with characteristics that both enable and challenge its role in the supply chain. In Section 3, traditional economic functions of cooperatives are discussed, such as risk sharing, bargaining, service provision, and marketing. Subsequently, the discussion will shift to new functions for cooperatives, such as vertical coordination in the supply chain, which has become more important in recent years (Section 4). The challenges these new functions bring for managing and organizing the cooperative are described and discussed in Section 5, focusing on financing, corporate governance, member commitment and member heterogeneity. Finally, in Section 6, I will conclude with a number of observations as to the strengths and weaknesses of cooperatives in the new market environment.

2. What are the ideal-type characteristics of a cooperative?

In this Section I will present the ideal characteristics of a farmer-owned cooperative. A cooperative is an organization established by agricultural producers for the purpose of supporting their economic well-being. Cooperatives vary in form and function from country to country and from product to product. This diversity is reflected in different objectives, economic dimensions, legal status and internal structures. In many countries, 'cooperative' has a distinct legal meaning, often different from other types of producer organizations.

A common definition of a cooperative is the one used by the International Cooperative Alliance (see: www.ica.coop): 'A cooperative is an autonomous association of persons united voluntarily to meet their common economic, social and cultural needs and aspirations through a jointly owned and democratically-controlled enterprise.'[45] In other words, a cooperative is both an association of members and an enterprise in which economic activities take place. It is in the association that social processes of decision-making, trust building and

---

[45] While this definition includes producer cooperatives where all production assets are collectively owned, in our chapter we focus on rural cooperatives where members own their own farm, and where the cooperative provides services (such as input provision, marketing and processing) to the members (or member firms).

informal communication take place, and it is through the enterprise that economic benefits for the members of the association are generated. The dual character of a cooperative has long been acknowledged in literature on agricultural cooperatives (see Draheim, 1955; van Dooren, 1982; Müncker, 1994).

Despite the spatial and temporal diversity found among cooperatives, they share a number of common characteristics[46], such as collective action, bottom up structure, member control, democratic decision-making, and an orientation towards member benefits. First, a cooperative is a form of collective action by persons with common interests. As agricultural production in most of the world is carried out by small farmers, there are good economic reasons for these farmers to pool their risks, collectively market their product, and to collectively buy or produce the farm inputs they need (sections 3 and 4 will discuss economic justifications). In other words, a cooperative groups individual farmers for the purpose of joint economic action. This economic function is performed by the jointly owned enterprise, which operates in competition with other member-owned and non-member-owned firms.

Second, a cooperative is established by the producers themselves. While support from outside parties such as the state or development NGOs is often important in the start-up phase, a cooperative is essentially a bottom-up type of organization. This also implies that cooperatives are member-owned and member-controlled.[47] Member ownership in the economic sense means that members supply the equity capital needed to run the business. Member ownership in the socio-psychological sense means that producers are convinced that they have control over the organization. Member control implies that members decide on the activities and investments of the cooperative. While the transaction relationship between the member and the cooperative is of an individual nature, the ownership relationship is of a collective nature. This means that individually members cannot exert control over the cooperative. Both control and ownership are collective. Related to the characteristics

---

[46] These characteristics are partly based on the seven cooperative principles as developed by the ICA, which are: (1) Voluntary and Open Membership; (2) Democratic Member Control; (3) Member Economic Participation; (4) Autonomy and Independence; (5) Education, Training and Information; (6) Cooperation among Cooperatives; and (7) Concern for Community (see: www.coop.org).

[47] In the economic literature on cooperatives, one often finds the following definition: a cooperative is an organization that is established for the benefit of the user, is user-owned and user-controlled (e.g., Barton, 1989).

of member ownership and member control is the voluntary nature of membership; producers can join or drop-out as long as they comply with membership regulations.

Third, a cooperative has a democratic decision-making structure. All members have a voice and at least one vote. Members jointly decide on the functions, strategies and investments of the cooperative. Details of the decision-making process may vary depending on the size the organization (e.g. large organizations often use a tiered decision-making structure) and the culture for political processes. There are differences as to the number of votes an individual member may hold. Despite the many differences, democratic decision-making is the standard.

Fourth, a cooperative is a user-oriented firm, not an investor-oriented or family-oriented firm. Users aim to make a profit (or a living) with their own business (their own farm), while the services of the cooperative support them in this aim. Applying a property rights perspective (Barzel, 1997), we can say that the users are the residual claimants. They benefit through use[48], and not through the investment they have in the cooperative. The more use a member makes of the services provided by the cooperative, the more he/she benefits from membership.

Fifth, a cooperative has a long-term perspective. As a cooperative is intrinsically tied to agricultural production, the time scope of the cooperative is directly related to the timeframe common in farming communities. As agricultural production requires long-term investments and is usually carried out in production units that are family-owned and passed on from generation to generation, a cooperative necessarily has a long-term perspective. This timeframe, both of the associated producers and the cooperative, supports the development of common norms and values (which in turn lead to reduced transaction costs).

Sixth, a cooperative is an association comprised of human beings. This implies that the cooperative is a social community, with social mechanisms like commitment, solidarity, informal communication, and particularly identity. Within a social community collective identities can develop. Collective identities are important for the community as they shape expectations about member behaviour and allow members to calculate the costs and benefits of their individual actions vis à vis the collective. Identity is a powerful source of meaning for social actors (Castells, 1997).

---

[48] Other terms for "use" found in the literature on cooperatives: patronage, or transaction relationship.

Producer organisations and market chains

An ideal cooperative combines all six characteristics mentioned above. As these characteristics together form a coherent system of attributes (Hendrikse and Veerman, 1997), a deviation may result in the cooperative functioning less efficiently, with a lower level of member commitment, or even threaten its sustainability. In real cooperatives, however, not all of these traits are well-developed, or (board) members have accepted a deviation from these idealised characteristics. The detailed organizational arrangements within a cooperative may differ from country to country, just as formal and informal institutions differ by country. A different institutional context may demand a gradual (but not a fundamental) departure from these ideal characteristics.

I will now discuss the functions of a cooperative: both traditional functions (in Section 3) and new functions (in Section 4) that are a result of the shift from mainly horizontal coordination in bargaining and risk sharing, to more vertical coordination, such as quality control and innovation in (international) agrifood supply chains.

3. Traditional economic functions

Traditionally, cooperatives were established for a number of reasons.[49] The bottom line is that the cooperative should enhance the prosperity of its members, the producers. In order to achieve this, the cooperative can carry out a number of economic functions such as collective marketing of farm products, collective purchase of inputs, sharing of risks, and collecting and transferring market information. In recent years, emphasis on functions of cooperatives has shifted towards so-called supply chain functions. As vertical relationships have become more important, cooperatives have strengthened or started engaging in activities like improving logistics, improving quality (control) and even helping members to develop new products.

Overcoming market failures has traditionally been one of the dominant economic arguments behind the existence of cooperatives. Markets do not always work efficiently or may even be absent. For instance, most agricultural production takes place in small production units (often households), while processing and marketing of farm products is done by relatively large firms. In these oligopolistic or

---

[49] Bosc *et al.* (2001) list the following five functions of rural producer organizations: economic, social, representation, information sharing, and coordination. Our focus is on economic functions, which also include the task of coordinating transactions between producers and their customers.

even monopolistic markets, farmers are likely to receive a lower price then they would under more competitive conditions. Cooperatives can address this imbalance in competitiveness by establishing countervailing power. Thus, enhancing the bargaining power of producers has always been one of the main functions of a cooperative as part of its objective to increase producer income.

The function of enhancing bargaining power implies a number of activities carried out by the cooperative, such as contract (and price) negotiations, physically collecting the farm products, and also making sure that the products are of homogeneous quality. The cooperative may apply particular quality standards. Without homogeneous quality, collective bargaining is more difficult as separate negotiation has to be done for each quality class. This is an argument in favour of member homogeneity, an issue we will discuss in Section 5.

Bargaining cooperatives have economic benefits that go beyond the members, as other buyers in the industry are also forced to pay at least the same price as that of the cooperative, to their suppliers. This is the competitive yardstick theory of cooperatives (Cotterill, 1984). In addition, the price negotiation process may be useful in itself as a form of price discovery in markets where there is uncertainty about market supply and demand conditions (Hueth and Marcoul, 2003). Moreover, bargaining cooperatives play an important role in ensuring contract reliability, and this can have significant consequences for market efficiency (Hueth and Marcoul, 2003; Bogetoft and Olesen, 2004).

Another function of the cooperative related to market failure is reducing information asymmetries. Well-functioning (i.e., competitive) markets require that all market participants have full information regarding demand and supply, both on the quality and the quantity of products to be traded. However, this situation of full information disclosure rarely exists in real life, and most transactions are characterized by one party having more information than the other. Such situations of asymmetrically distributed information between buyers and sellers can lead to a lack of trade, even when trade would be beneficial to both (Akerlof, 1970). Asymmetric information is quite common in the relationship between the farmer and his transaction partner, such as the seller of inputs or the buyer of farm products. For instance, farmers individually do not have the knowledge and equipment to assess the quality of inputs such as fertilizers and animal feed. By setting up a cooperative that buys and tests the inputs, producers not only strengthen their bargaining position but also share the costs of quality assessment equipment. In many cases these supply cooperatives developed into input manufacturers themselves, particularly in the animal feed industry.

Another example of how cooperatives can solve the problem of information asymmetry can be found in the provision of credit. For financial institutions, often located outside the rural area, it may be difficult to assess the credit worthiness of (small) farmers. As a result these institutions either refuse to provide credit or provide it at high interest rates. Rural credit cooperatives have solved this information asymmetry problem by using the detailed information available to people who need credit, the farmers themselves. The capability of a cooperative to efficiently gather and use creditor information was one of the main reasons for the rapid growth of rural credit cooperatives in Europe and North America in the 19th and 20th centuries (Bonus, 1986). Another advantage of rural credit cooperatives over commercial banks is the economic and social incentives that exist among members to repay the loan.

Cooperatives have also solved the asymmetric information problem in marketing of farm products. When buyers have better information on consumer demand in final markets, they can use this information advantage when negotiating price and delivery conditions. One approach to this information problem is for a group of farmers to hire a sales agent. This route has been taken in many countries, where bargaining cooperatives not only negotiate with potential buyers, but also gather and process market information. However, employing a sales agent may involve other informational problems. When both production and demand are volatile, for instance due to weather conditions, it is hard to assess whether the sales agent has put in the effort that was contractually agreed. This double information problem, on the market conditions and on the effort of the sales agent, can be solved by collectively organizing an auction. The famous Dutch flower auctions are examples of cooperatives that have solved information asymmetry between producer and buyers. By centralizing all supply and demand and by using the auction clock as a price determination mechanism, the market becomes transparent, efficient and fair. In addition to the efficiency obtained in the market, efficiency in production is also enhanced because producers can fully focus on and specialize in production activities.

Another market failure that cooperatives might address is the lack of processing facilities. Outside financers may not be willing to invest in a processing plant because the return on investment is too low. Unwillingness to invest in processing facilities may also be the result of the transaction specificity of the investment. As earning back the investment depends on continuity of supply from a specific group of producers, the processing firm is dependent on these producers. This is the asset specificity argument of Transaction Cost Economics (Williamson, 1985). If party A to a transaction with party B has to make an investment that is specific to the transaction[50], A becomes dependent on B. This dependency may turn out to be costly when B decides to discontinue the contract before A has earned back his investment. B may even take advantage of A's dependency situation by renegotiating the contract. This threat of loss of investment, including the threat of opportunistic behaviour by B, may be sufficient justification for A not to make the initial investment. In other words, no efficient transaction will take place. A cooperative that invests in processing the products of its members solves this transaction cost problem. This argument has often been claimed as the main justification for the existence of dairy cooperatives (Staatz, 1987).

Cooperatives can be a vehicle for sharing investments, such as farm equipment. Machinery associations or farm equipment cooperatives, in which farmers collectively own tractors, harvest combines or other machines, have developed all over the world. These cooperatives can particularly be found in situations where farm machines have a high purchase price and where the capacity of the farm machine exceeds the size of an individual farm (Kuhn, 1994).

Developing and exchanging technical knowledge can also be a function of the cooperative. The simplest form of knowledge sharing cooperative is an association of farmers who regularly discuss technical (or managerial) problems with one another and visit each other's farm. A more elaborate form of this function is a cooperative that contracts agricultural research. Fulton (2005) mentions producer-funded research programs for pulse and canola in Canada. However, this type of contract research for particular farmer groups is usually organized through farmer unions. Bosc *et al.* (2001) argue that the role of cooperatives in the creation and diffusion of technical innovations will become more important in the near future because national research and extension

---

[50] Transaction specificity means that the investment has lower value in alternative uses.

systems in developing countries are often non-productive and inefficient regarding farmers' needs, and also due to ongoing reforms of state institutions.

Finally, a cooperative is an instrument for sharing risk. The more vulnerable producers are to the vagaries of nature or the whims of the market, the more they need risk reduction strategies. Many examples of cooperatives with a risk sharing function exist, such as mutual insurance companies[51], farm help associations to share the risks of illness, and marketing associations using a pool system to share the price risk in selling farm products.

4. New economic functions: from horizontal to vertical coordination

So far, we have focused on the more traditional functions of agricultural cooperatives. The common characteristic of these functions is that they reduce costs and solve market failures. By acting collectively, producers save on the costs of processing and marketing, purchasing inputs, obtaining credit, reducing risk, purchasing farm equipment, and obtaining technical information. In addition, producers strengthen their bargaining power by collectively negotiating with suppliers or buyers. As we have written above, the last decade has seen a shift in the main functions of most cooperatives, due to the increasing importance of vertical supply chain coordination. While traditional functions continue to be important for all cooperatives, a number of new functions have been taken up by those cooperatives that want to strengthen their position in the supply chain. These new tasks relate to improving and guaranteeing quality, enhancing logistic efficiency, reinforcing information exchange, and strengthening innovation. Instead of cost reduction and strengthening market position, these new tasks are more targeted at value creation, either at farm level or at the whole supply chain level (e.g. branded products).

One of these new functions relates to information being unequally distributed in the supply chain. A cooperative can address information asymmetry regarding product quality. Producers have better information on product quality, particularly on quality characteristics that are directly related to on-farm production methods. These quality characteristics, such as organically production, are hard for the buyer to measure. For this reason, buyers may not be willing to buy the product, or will only

---

[51] Most mutual insurance companies started as mutual fire insurance for farmers.

do so at a low price. This, again, is the classical information asymmetry problem posed by Akerlof (1970), which leads to a loss of welfare. When the buyer of these high quality farm products is a farmer-owned cooperative, a major part of the information problem between seller (farmer) and buyer (coop) is solved, as there is no conflict of interest between the two. In addition, cooperatives may have lower costs in guaranteeing compliance of their suppliers (i.e. their members) to quality requirements. Economic as well as social incentives keep members of the cooperative from behaving opportunistically.

Still, there may be an information problem regarding product quality in the transaction between the cooperative and its customers. Customers may not be able (or only at high costs) to measure the quality of the product, particularly its so-called credence attributes.[52] This information problem can be solved by having the cooperative establish a reputation, often embodied in a trade or brand name that functions as a signal or as a credible commitment (Williamson, 1985) to its buyers indicating that it has a continuing interest in supplying the agreed quality. As farm products are sold repetitively, reputation is a very important instrument for reducing transaction costs related to information asymmetry. Applying particular (private and public) quality standards may also signal to customers that the product is of good quality. A cooperative can also help processors to obtain information on the quality of their supplies. They need this information to optimize their processing activities, both in terms of maintaining quality and operating equipment as efficiently as possible. [53]

In some situations it may be easier for a cooperative to establish a brand name than it would be for other chain actors. In those cases where the quality of the branded product is crucially dependent on the effort of the producer (e.g. in the case of regional specialties), no other chain actor is willing to invest in a brand name. Cooperatives are in the best position to invest in this brand, not only because members jointly make the investment, but also because there is no conflict of

---

[52] Credence attributes of a product are those attributes a buyer (e.g. a consumer) cannot measure. In accepting these attributes, the buyer relies on the credentials (or reputation) of the seller (e.g. a producer). An example is the organic nature of a food product.
[53] Hueth and Marcoul (2003) suggest that one of the reasons for the formation of bargaining cooperatives in the US is that they have been able to share production information with processors that would otherwise be costly to obtain. Bijman and Hendrikse (2003) argue that bargaining associations in the Dutch fresh produce industry have been set up to improve coordination between growers and their customers (traders or retailers). Bogetoft and Olesen (2004) describe the coordination role of several cooperatives in the Danish agrifood industry.

interest between the producers and the cooperative, so there is no risk of opportunistic behaviour on the part of the producers (Raynaud *et al.*, 2005). Besides investment in a brand (i.e., in a specific reputation), quality products also often require additional investment in production. Farmers may not be willing to make these investments, if they do not have guaranteed market access or if they cannot fully rely on their customer(s) to continue marketing these high quality products. The cooperative can also solve this problem of market failure, as it can guarantee that it will do everything needed to market the (special) products of its members. Due to investment risk, both on the side of the producer and on the side of the brand owner, most organic products and products of regional origin are marketed through cooperatives (Verhaegen en Vanhuylenbroeck, 2002).

Cooperatives are also faced with the need to set up a quality assurance (and traceability) system. Customers increasingly demand (and in some countries legislation even requires) the exchange of detailed information regarding the processes applied at different stages of the supply chain. A cooperative, being at the interface between production and trade/processing, is in a good position to set up such a quality control system. Being a producer-controlled organization, the cooperative can relatively easily obtain information from its members and transfer this information to other chain actors. Processing cooperatives in particular, have been active in establishing quality control systems. The dairy industry, for instance, is known for the leading position of farmer-owned cooperatives in developing and applying quality assurance systems.

A cooperative can also support its members by providing information on the qualitative and quantitative market demands. First it can exchange general information on market trends and more specific information on customer requirements. Second it can help its members to comply with the quality requirements demanded of customers, by supplying certain inputs; giving training and technical assistance; and by helping them to learn from each other's experiences (in the case of farmer study groups). In addition, members may more easily accept visits by quality supervisors that work for the cooperative than those that work for a commercial customer. The latter is considered an adversary (in the vertical competition), who should be given as little information as possible.

In coordinating the activities of supply chain partners, the cooperative can carry out the following activities. The cooperative may organize the standardization of packaging (such as crates) and it may organize logistic processes such as storage, sorting and grading, and transport. In addition, the cooperative may coordinate strategic issues, by aligning the interests of different chain participants. Take for instance the case of product innovation. While producers must comply with basic quality requirements demanded by their customers in order to maintain market access, they can also develop new products. Nowadays product innovation in agriculture will only be economically successful when simultaneous effort is put into innovation at production and marketing levels of the chain. New products often require innovative marketing approaches. The cooperative occupies a central position in the supply chain for coordinating complementary adjustments in production and marketing.

As coordination requires information exchange and decision-making, information on consumer demands for food quality must be passed on to producers, while information on the quality of the products needs to reach customers. The cooperative, being a central authority on behalf of the producers, can organize this information exchange. The cooperative has certain decision-making rights regarding member activities, so it can apply specific organizational governance mechanisms to obtain the coordination that is necessary (Royer, 1995). Verhaegen and Van Huylenbroeck (2002) in their Belgium study, found that producers of quality food products often set up cooperatives in order to coordinate both horizontally among the producers (to get homogeneous quality products) and vertically between producers and customers (in order to align the supply and demand of high quality products).

These shifts in functions, from horizontal to vertical collaboration and coordination, pose new challenges for the management and organization of cooperatives, as they have implications for investments made by the cooperative, the distribution of decision-making rights, and member commitment. I now turn to the question of how new supply chain functions affect the organization and management of cooperatives.

5. Organizational and managerial challenges of cooperatives

Cooperatives have always been prone to internal challenges resulting from specific organizational characteristics like collective decision-making and member-ownership. For instance, cooperatives have to deal with the trade-off between scale economies, which are part and parcel of a large membership, and high member commitment resulting from a small membership. However, if cooperative seek to enhance vertical coordination between producers and customers, new challenges emerge.[54] In this section we focus on four issues related to organization and management that have to be dealt with by cooperatives seeking to strengthen their role in the supply chain: financing, corporate governance, member commitment, and membership heterogeneity.

5.1 Financing the cooperative

When producers have to comply with the quality requirements of their domestic and foreign customers, they often need to make investments to improve production processes as well as in storage, sorting, packaging, processing and transporting processes. Often hygiene and cooling requirements demand the purchase of new equipment, the building of new facilities, and the upgrading of quality measurement procedures. In addition, cooperatives that want to become (or remain) preferred suppliers of major retail customers in developed countries continuously invest in innovation, such as new products, new packaging, or new marketing programs.

All of these new tasks, which are needed in order to supply highly critical customers, require additional investments. While production is mainly the responsibility of individual producers, improving post-harvest activities is the task of the cooperative and thus a collective responsibility. The main question for cooperatives is where to obtain this capital. In developed countries, equity capital traditionally comes from the members, either directly in the form of a membership fee or indirectly as retained earnings. In developing countries, external stakeholders often provide the initial capital. However, even without new demands related to being a partner in the supply chain, cooperatives have had difficulties in obtaining sufficient risk capital. A member of a cooperative is always asking himself: do I invest in my own farm or in the cooperative? Let us discuss this issue of financing the cooperative in more detail, using an organizational economics perspective.

Cooperatives are collectively owned by the farmer-members. Collective ownership means that control rights and return rights are not assigned to any individual member, but are held jointly by all members

---

[54] We do not discuss the problems of state lead cooperatives. These coops were known for their low membership commitment. With the restructuring of national economies and the retreat of the state from direct economic activities, these cooperatives are struggling with their new role, and with convincing current and potential members that they have truly reorganized into bottomup type of organizations.

together. Most cooperatives have no or limited options for trading property rights. The economic theory on the importance of property rights (e.g., Barzel, 1997) has shown that ill-defined and non-tradable property rights lead to inefficient decisions. If it is not clear who owns an asset and/or the asset cannot be traded, economic agents will not invest in the asset. Cook (1995) has distinguished three investment related efficiency problems in cooperatives: the free rider problem, the horizon problem, and the portfolio problem. The free rider problem occurs if someone other than the investor benefits most from the investment. For instance, non-member farmers can profit from the marketing activities of a cooperative for generic products, without sharing the costs; or new members can profit from joining the cooperative without having to pay an entrance fee. The horizon problem arises if an investment has to be made that only pays off in the long term. Particularly older members are unwilling to invest in assets that generate (most of the) income after they have retired. The portfolio problem refers to the situation wherein individual members cannot adjust their share (i.e., their investment) in the cooperative to match their personal risk preference. If risk is higher than some members threshold, then these members will not invest. Also, risk adverse members may influence the management to abstain from investments that are beneficial to the cooperative as a whole.

The main efficiency problem of collective ownership of a cooperative that is a partner in a supply chain is that members are unwilling to provide the equity capital needed for making necessary investments in quality improvement, innovation, or marketing. Particularly investments in innovation and marketing (such as establishing a brand name) entail high risk. As debt capital is not the most appropriate (because it is more expensive) source of capital for such investments, members are the most likely source.

Both the theoretical and the empirical literature suggest that the incentives for members to invest in cooperatives would be enhanced if property rights were better defined. One way of doing this is to individualize part of the collectively owned capital, as has been done by several dairy cooperatives in New Zealand, Australia and the Netherlands (van Bekkum, 2001). Another option is to introduce delivery rights that can be traded and appreciated. Cook and Iliopoulos (2000), in a study of American cooperatives, found that members of cooperatives with a closed membership policy, that use marketing agreements and have transferable and appreciable delivery rights, are more willing to invest in their cooperative than members of cooperatives with open membership, without marketing agreements and with non-transferable and non-appreciable shares. In North America, cooperatives that have a closed membership policy, use marketing agreements, and have transferable and appreciable delivery rights are called 'new generation cooperatives' (Harris *et al.*, 1996).

In order to gain additional capital many cooperatives in developed countries have changed their organization structure, either to encourage their members to supply additional equity capital or to invite outside investors to provide equity capital for the cooperative. Chaddad and Cook (2004) present a typology of organizational models that can be used to solve the investment problem. Although these authors have built their typology mainly on the basis of cases from developed countries (mainly Europe and North America), the theoretical basis allows for a broader (geographical) scope. The theoretical basis of the typology is the organizational economics theory on ownership, where ownership is defined as the combination of residual control rights and residual claim rights (Hansmann, 1996). The authors distinguish seven discrete ownership models (see Figure 1), ranging from traditional cooperatives at one extreme to investor-owned firms (IOF) at the other. In between these extremes are five models that in discriminative ways combine the ownership structure of traditional cooperatives (where member/ users have full residual control and claim rights) and of IOFs (where investors hold all ownership rights). Chaddad and Cook emphasize that the five non-traditional models can be used by cooperatives to ameliorate perceived financial constraints (while retaining a member-owned organization).

5.2 Corporate governance

Governance of cooperatives has several specific characteristics that may lead to challenges if the cooperative wants to strengthen its position in the supply chain. First, as stated above, a cooperative is a member-owned and member-controlled firm. The producer-members are not only the owners in the sense of residual claimants, but also control the cooperative. All major decisions have to be taken by the producer-members themselves or their representatives (usually a board of directors). This also means that all activities and all investments of the cooperative should be in the interests of the producer-members. Thus, cooperatives have internal organizational structures that recognize

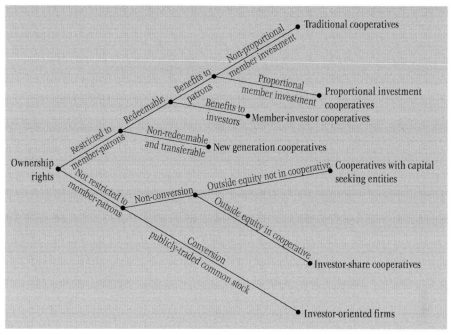

Figure 1. Ownership rights typology of cooperative models (Chaddad and Cook, 2004).

member rights and interests.[55] Cooperatives that want to strengthen their position in the supply chain, for instance by complying with customer requirements, will become relatively more customer oriented. This means that in (strategic) decisions of the cooperative, customer interests and member interests will both be taken into account. It is the task of the professional management to ensure that customer interests are sufficiently served, even when they conflict with (short term) member interests. It can be expected that cooperatives aiming for supply chain integration will experience more difficult decision-making as member interests are no longer the only (or even the main) guiding principle.

Second, a cooperative is characterized by collective decision-making. While collective decision-making has clear advantages for the quality of decisions, it also has disadvantages in terms of speed. As the majority of

---

[55] The influence of the individual member on cooperative decisions is limited. Member influence is usually confined to electing and monitoring the board of directors. The board of directors takes most decisions, while the execution of these decisions is delegated to professional management.

Producer organisations and market chains

the membership must approve an investment, decision-making tends to take more time in a democratic structure as compared to an autocratic structure (such as an IOF). In addition, collective decision-making processes tend to avoid high-risk investments, which are sometimes needed for innovation and marketing purposes. As a result, cooperatives may have difficulty in taking decisions regarding investments required in order to become and remain a preferred supplier for large domestic and foreign retailers.

Third, when cooperatives become larger and more customer-oriented, the distance between the association part and the firm part of the cooperative increases. Particularly the professional managers of the firm want sufficient room for entrepreneurial initiatives. This means that the relationship between the board of directors (as representatives of the membership) and the management may change from a situation where the board takes strategic decisions and management takes care of execution, to a situation where management takes the strategic decisions and the board's main function is to monitoring ex post (as well as explaining and defending these decisions to the members). An organizational option for giving the management more freedom of initiative is the introduction a formal (i.e., legal) distinction between association and firm. In the Netherlands, several large agricultural cooperatives have reorganized their corporate governance structure along these lines. The cooperative firm has been transformed into a limited (liability) company (Ltd) or a public limited company (Plc), and the association has become a holding company, usually having a 100% share of the (public) limited company.[56] This implies redefining the allocation of authority between the board of directors and (professional) management by giving the latter more authority in operational and even strategic matters. It also implies a larger administrative distance between members of the association and the firm. Reasons for these changes in corporate governance structure were reducing liability, spreading risks, and a more formal distinction between the association and the commercial activities of the cooperative firm (Van der Sangen, 2001).

---

[56] In Ireland several dairy cooperatives have introduced this legal separation as a first step to obtain stock listing. Once the separation has been formalized, the association can sell part of its shares to public or private shareholders.

5.3 Member commitment

Cooperatives becoming more customer-oriented can no longer focus only on member interests. This may result in reduced member commitment, causing serious efficiency problems for the cooperatives. Waning member commitment is a result of a perceived lack of connection between member efforts and cooperative success, combined with the inability of the cooperative to differentiate itself from other organizations or business entities. According to Fulton (1999: 418), 'member commitment is critical because it is a measure of how well a co-op is able to differentiate itself from an investor-owned firm (IOF).'

Member commitment in cooperatives is important for several reasons. First, commitment is needed for financing the cooperative as members are the main source of equity capital. Low member commitment will probably lead to low willingness to invest in the cooperative, which is a problem for increasing customer orientation as we have discussed above. Second, commitment is also needed for efficient coordination between producers and their cooperative: members should provide information on product quality and quantity they supply in order for the cooperative to properly and efficiently process and market members' products. Third, commitment is needed for the sustainability of the cooperative: members not committed to their cooperative may easily switch to other business partners, thereby jeopardizing the very existence of the cooperative (particularly cooperatives that were established in order to have economies of scale). Fourth, commitment is required for efficient and effective decision-making and control. In a voluntary, collective organization, members need to be involved in decision-making. Low commitment leads to low willingness to engage in decision-making processes, and thus to inefficient control over the management of the cooperative. Low commitment leads to higher influence costs i.e., members trying to influence decision-making for individual (compared to collective) benefits. Fifth, commitment is needed for building and maintaining common norms and values. These common norms and values are needed in order to keep transaction costs low both among the members and between members and cooperative. Finally, commitment is needed so that members abstain from opportunistic behaviour: low commitment may lead to opportunistic (or free rider) behaviour by individual members in their (transactional) relationship with the cooperative.

Cooperatives that aim at being an active partner in a (international) supply chain may have more difficulty to keep members committed, because distant markets may be more volatile, less transparent and may entail higher risks. At the same time these markets may require members to provide additional equity capital and to put in additional effort to improve quality. What are the options for cooperatives in terms of maintaining member commitment? On the basis of a literature study, Kroft (2006) found that the following factors positively influence member commitment: economic benefit, opportunities for member influence in decision-making process, legitimacy of the board, strength of cooperative ideology, communication between the cooperative and its members, and member trust in the cooperative.

5.4 Member heterogeneity

Member heterogeneity is a problem where members deliver heterogeneous products to the cooperative while homogeneous products are demanded by the customers. As not all members are equally able to reduce variability in product quality, measures by the cooperative to reduce this variability will trigger different interests within the membership. The more important vertical coordination is in the supply chain, the more problematic variability in product quality will be. Moreover, as the cooperative becomes more customer-oriented, it may need to invest in product innovations that cannot favour all members equally.

Membership homogeneity is important for building and maintaining a collective identity in the cooperative. This identity is, in turn, important as it results in trust, for instance between the membership and the management. Borgen (2001) found that the stronger the members' identification to the collective organization, the more they trust the benevolence of cooperative management. As there is a significant and increasing information asymmetry between members and management on market prices, customer behaviour etc., it is crucial for the smooth working of the cooperative that members trust the capability and intentions of the management.

Social and functional heterogeneity pose some serious challenges for cooperatives. Decision-making may become more laborious, coordination between member-producers and the cooperative may become more difficult, member commitment may decrease and member willingness to provide equity capital may be reduced. When the number of separate activities carried out by the cooperative increases, sub-group lobby or pressure is likely to emerge resulting in a rise in influence costs. In addition, the horizon and free rider problems can be intensified and membership commitment is likely to decline. Some members could decide to leave the cooperative and start a new group. In sum, membership heterogeneity will affect the efficiency of the cooperative.

On the basis of an extensive review of the economic and organization literature regarding membership heterogeneity in agricultural cooperatives, Bijman (2005) argues that cooperatives should seek homogeneous membership in order to improve organizational efficiency. Homogeneous membership leads to higher member commitment, lower decision-making costs, lower influence costs incurred by the cooperative, lower agency costs, a more focused strategy by the cooperative, stronger member incentives to provide equity capital, and lower coordination costs in the member-coop relationship.

6. Conclusions

Analysis in this chapter is mainly based on the experiences of cooperatives in developed countries. However, we are convinced that these issues are also relevant for cooperatives and other POs in developing countries: most of the internal challenges are valid for member-owned firms all over the world. But more importantly, cooperatives in developing countries increasingly face the same problems as coops in developed countries as they become partners in (international) supply chain and as they strengthen vertical coordination among supply chain actors.

The primary function of cooperatives in developed and developing countries alike is shifting from mainly horizontal towards more vertical coordination. This implies taking up new tasks such as performing quality control, setting up quality assurance systems and introducing new (private) grades and standards. In addition, exchanging information has become more important for cooperatives, both information about market requirements from customers to producers and information about product characteristics from producers to customers. Setting up a traceability system may be a good starting point for cooperatives to strengthen their role in vertical coordination. Helping members to improve quality, by providing compliance information and training may be one of the main new functions of the cooperative. Besides quality improvement, standardization, and logistics, cooperatives can enhance their marketing effort by establishing some kind of reputation. This requires a long-term perspective and often significant investments. As these marketing efforts entail investments in transaction-specific assets, cooperatives may be advised to enter into contracts, both with customers and with their own members.

As we have argued in Section 5, these new supply chain functions may pose additional challenges for the organization and management of the cooperative. When customer relations and marketing expertise becomes relatively more important, professional (marketing) managers may be needed to run the cooperative. However, having professional management in a member-owned firm requires a board that is sufficiently equipped to control the management. Thus, improving the knowledge and skills of the board members is needed to maintain an efficiently managed cooperative. The issue of homogeneity of products is becoming more important when cooperatives face more demanding customers (such as supermarkets). Thus, cooperatives should on the one hand help members to improve product homogeneity, and on the other hand try to reduce membership heterogeneity. However, if reducing membership heterogeneity may result in a smaller firm, thus losing economies of scale and bargaining power, the cooperative may try to set up collaboration with other cooperatives, bilaterally or multilaterally in a federation, in order to generate the countervailing power needed to maintain vertical competitiveness. When cooperatives impose stricter quality regulations on their members, communicating with members is even more important. Explaining why such regulations are necessary and how individual members can comply with them remains a very important task of the collective organization.

Finally, cooperatives must find the capital needed to set up marketing programs, quality control programs and innovation programs required to make them valued partners in the supply chain. Here a major difference between cooperatives in developed and developing countries exists. In developing countries, producer-members generally do not have capital available. NGO and state support may be a solution for financial constraints. This, however, can only be a viable solution when members remain the owners of the collective firm, both in the legal and psychological sense of ownership.

References

Akerlof, G.A., 1970. The market for 'lemons': qualitative uncertainty and the market mechanism. Quarterly Journal of Economics, 84: 488-500.

Barton, D.G., 1989. What is a cooperative? Cooperatives in agriculture. D. W. Cobia. Englewood Cliffs, NJ, Prentice Hall: 1-20.

Barzel, Y., 1997. Economic Analysis of Property Rights. Second Edition. Cambridge, UK, Cambridge University Press.

Bekkum, O.-F. van, 2001. Cooperative Models and Farm Policy Reform. Exploring Patterns in Structure-Strategy Matches of Dairy Cooperatives in Protected vs. Liberated Markets, Assen: Van Gorcum.

Bijman, J. and G. Hendrikse, 2003. Co-operatives in chains: institutional restructuring in the Dutch fruit and vegetables industry. Journal on Chains and Network Science, 3(2): 95-107.

Bijman, J., 2005. Cooperatives and heterogeneous membership: eight propositions for improving organizational efficiency, Paper presented at the EMNet-Conference Budapest, Hungary, September 15 - 17, 2005.

Bogetoft, P. and H.B. Olesen, 2004. Design of production contracts. Lessons from theory and agriculture. Copenhagen, Denmark, Copenhagen Business School Press.

Bonus, H., 1986. The cooperative association as a business enterprise: a study in the economics of transactions. Journal of Institutional and Theoretical Economics / Zeitschrift für die gesamte Staatswissenschaft, 142(2): 310-339.

Borgen, S.O., 2001. Identification as trust-generating mechanism in cooperatives. Annals of Public and Cooperative Economics, 72(2): 209-228.

Bosc, P.-M., D. Eychenne, K. Hussein, B. Losch, M.-R. Mercoiret, P. Rondot and S. Macintosh-Walker, 2001. The Role of Rural Producers Organisations (RPOs) in the World Bank Rural Development Strategy. Background Study. Washington, DC, World Bank.

Castells, M., 1997. The power of identity. Oxford: Blackwell Publishers.

Chaddad, F.R. and M.L. Cook, 2004. Understanding new cooperative models: An ownership-control rights typology. Review of Agricultural Economics, 26(3): 348-360.

Cook, M.L., 1995. The Future of U.S. Agricultural Cooperatives: A Neo-Institutional Approach. American Journal of Agricultural Economics, 77(December): 1153-1159.

Cook, M.L. and C. Iliopoulos. 2000. Ill-Defined Property Rights in Collective Action: The Case with US Agricultural Cooperatives, in C. Ménard, ed. Institutions, Contracts and Organizations, Edward Elgar, Cheltenham, U.K., 335–348.

Cotterill, R.W., 1984. The Competitive Yardstick School of Cooperative Thought, American Cooperation, pp. 41-56, American Institute of Cooperation, Washington DC.

Draheim, G., 1955. Die Genossenschaft als Unternehmungstyp. Göttingen.

Dooren, J.P. van, 1982. Co-operatives for developing countries. Amsterdam, Royal Tropical Institute.

Fulton, M., 1999. Cooperatives and Member Commitment. Finnish Journal of Business Economics 48(4): 418-37.

Fulton, M., 2005. Producer Associations: the international experience. China's Agricultural and Rural Development in the Early 21st Century. B.H. Sonntag, J. Huang, S. Rozelle and J.H. Skerritt. Canberra, Australian Centre for International Agricultural Research: 174-196.

Hansmann, H., 1996. The ownership of enterprise. Cambridge, MA / London, The Belknap Press of Harvard University Press.

Harris, A., B. Stefanson, and M. Fulton, 1996. New Generation Cooperatives and Cooperative Theory. Journal of Cooperatives, 11: 152-22.

Hendrikse, G.W.J. and C.P. Veerman, 1997. Marketing Cooperatives as a System of Attributes. Strategies and Structures in the Agro-Food Industries. J. Nilsson and G. v. Dijk. Assen, Van Gorcum: 111-130.

Henson, S., and T. Reardon. 2005. Private Agri-food Standards: Implications for Food Policy and the Agri-food Systems. Food Policy (30) 3. 241 253.

Hueth, B. and P. Marcoul, 2003. An Essay on Cooperative Bargaining in U.S. Agricultural Markets. Journal of Agricultural & Food Industrial Organization, 1(1): Article 10.

Kroft, C., 2006. Hoe betrokken ben jij? Onderzoek naar ledenbetrokkenheid en ledencommunicatie binnen Campina, Wageningen: Wageningen University (MSc Thesis Business Administration).

Kuhn, J., 1994. Machinery Associations and Machinery Rings in Agriculture, In: E. Dülfer (ed.) International Handbook of Cooperative Organizations. Göttingen, Vandenhoeck & Ruprecht, pp. 574-579.

Müncker, H., 1994. Organizational Structure of Co-operative Societies. In: E. Dülfer (ed.) International Handbook of Cooperative Organizations. Göttingen, Vandenhoeck & Ruprecht, pp. 656-661.

Raynaud, E., L. Sauvee and E. Valceschini, 2005. Alignment between Quality Enforcement Devices and Governance Structures in the Agro-food Vertical Chains. Journal of Management and Governance, 9: 47-77.

Reardon, T., C.P. Timmer, C.B. Barrett, and J. Berdegue, 2003. The Rise of Supermarkets in Africa, Asia, and Latin America. American Journal of Agricultural Economics, 85(5): 1140-1146.

Royer, J.S., 1995. Potential for Cooperative Involvement in Vertical Coordination and Value-Added Activities. Agribusiness; An International Journal, 11(5): 473-481.

Sangen, G.J.H. van der, 2001. Corporate governance bij coöperaties. Ontwikkelingen op het gebied van de structuur en de inrichting van de coöperatie. Agrarisch Recht, 61(7/8): 435-442.

Shepherd, A.W., 2005. The implications of supermarket development for horticultural farmers and traditional marketing systems in Asia. Rome: FAO (Agricultural Management, Marketing and Finance Service).

Staatz, J.M., 1987. The Structural Characteristics of Farmer Cooperatives and Their Behavioral Consequences. Cooperative Theory: New Approaches. J. S. Royer. Washington, DC, USDA, Agricultural Cooperative Service: 33-60.

Verhaegen, I. and G. Van Huylenbroeck, 2002. Hybrid governance structures for quality farm products. A transaction cost perspective. Aachen, Shaker Verlag.

Williamson, O.E., 1985. The Economic Institutions of Capitalism. Firms, Markets, Relational Contracting. New York, Free Press.

Chain empowerment: supporting African smallholders to develop markets[57]

Lucian Peppelenbos and Hugo Verkuijl

Value chains often exclude the poor. This chapter explores how smallholder farmers in Africa can benefit from value chains. The authors present a framework that helps producer organizations to reflect on their position in the value chain, and they illustrate the framework using a case study of shea butter production by rural women in Mali.

1. Introduction

Value chain development has become a key concept in debates on rural development and poverty alleviation in low-income countries. The chain approach holds promise in terms of secured markets and value-

---

[57] This chapter is largely based on KIT / Faida MaLi / IIRR (2006), *Chain Empowerment: supporting African farmers to develop markets*. We are grateful to all persons and organizations who contributed to this book. This chapter is as much a collective product as the book itself.

addition for rural communities. But at the same time it is recognized that value chains often exclude the poor. In this chapter we explore how smallholder farmers in Africa can benefit from value chains. In what ways can producer organizations participate in value chains to improve their members' livelihoods? How can they develop a degree of control over governance of the chain? To address these questions, we present a framework that helps producer organizations to reflect upon their position in the value chain. The framework will be illustrated with a real-life case study of shea butter production by rural women in Mali. The chapter concludes by defining basic strategies for empowering producer organizations in value chains. We start however by exploring the economic challenges facing smallholder farmers in Africa.

2. Economic globalization and African smallholders

Rapid economic changes are challenging the ability of African smallholders to supply their products to the market. These changes include: (a) the globalization and liberalization of agri-food markets; (b) the emergence of tightly coordinated supply chains; and (c) the decline of government support for agriculture.

Over the past 20 years, new trade policies have liberalized and integrated markets around the world. Some farmers have benefited. But many farmers in developing countries have seen their incomes fall. Their terms of trade have declined steadily as prices of agricultural commodities have fallen compared to manufacturing. Tightly coordinated supply networks have emerged. Buyers and sellers now sign contracts to produce specialized products, grown to strict specifications and packaged in a particular way. The buyer may negotiate a contract directly with a grower, rather than buying through a trader. This new organization of supply chains is unfamiliar territory for many African farmers. It is very different from the conventional, arm's-length trade in bulk commodities such as maize or wheat, which may involve many intermediaries, and where the buyer may not know who the producer is.

The emergence of value chains is accompanied by market concentration, with a small number of transnational companies dominating large parts of the agri-food system. In Africa, supermarket chains such as Shoprite, Uchumi and Nakumatt are beginning to dominate food retailing. They account for 30% of the food retail trade in Kenya, and 55% in South Africa (FAO, 2003). Supermarkets enforce stringent standards for the produce they buy. They want beans to be of a uniform length, mangoes to ripen at exactly the right time, and bananas to be free of bruises. Supermarkets pay growers attractive prices to ensure that they get produce of the right quality. But the rules are hard for smallholder farmers to comply with as they lack the right technology and management skills. As such, smallholders are denied access to a lucrative market.

Finally, structural adjustment programmes have meant that developing country governments have significantly reduced support to farming communities. Investments in rural infrastructure, input subsidies, marketing schemes, and services such as extension and research have all declined. In the past, most African governments provided services to farmers through commodity marketing boards and state-supported cooperatives. The decline of these institutions has left the majority of smallholder farmers less organized than before. They are now trying to increase production in the face of reduced inputs and declining prices. This increases the supply of low-quality goods onto the market, which further suppresses prices. This phenomenon is known as 'Cochran's treadmill' (Cochran, 1979): more farmers supply more products into a market where prices are steadily falling, natural resources are being degraded and poorly managed farming systems are spreading onto marginal land.

Most of the current debate on poverty alleviation focuses on increasing farmer incomes by improving access to new markets. In this chapter we argue that market development alone is not enough to improve rural enterprises and livelihoods. The latest Economic Report on Africa shows that despite record economic growth, poverty is actually getting worse, particularly in the countryside (UNECA, 2005). This is because economic growth has had limited effect on local value chains, the generation of jobs, and building up of better economic institutions. For development to take place, various actors in the value chain must invest in a coordinated way. Government investments in rural infrastructure are profitable only if rural producers and their organizations also invest in increased production, local businesses invest in processing and distribution, service providers invest in new technology, and so on. If these complementary efforts are not well-coordinated, a so-called 'equilibrium of underdevelopment' may occur (Stockbridge *et al.*, 2003; Hoff, 2001).

Producer organizations play a key role in better economic coordination in the countryside. They reduce transaction costs by coordinating the business operations of many smallholders. They can train and support member farmers to understand market demands and to supply the required volumes of quality product at the right time. Farmer organizations can build up relationships with various chain actors and create commitment to cooperate on mutually beneficial actions and investments. In this chapter we focus on these roles for farmer organizations in value chains. For this purpose we present a framework to help farmer organizations reflect upon their position in the value chain.

3. Farmers' participation in value chains

Smallholder farmers can participate in value chains in many different ways. These forms of participation can be assessed according to two broad dimensions: (a) the types of activities that farmers undertake in the chain; and (b) the involvement of farmers in the chain governance (Peppelenbos, 2005; KIT *et al.*, 2006).

Chain activities. Farmers can choose to focus only on production, land preparation, growing the crop, and harvesting when the crop is mature. But they may also be involved in other chain activities, such as drying their crop, sorting and grading, processing, transporting and trading. Being involved in various activities in the chain is known as vertical integration (Figure 1).

Chain governance. Farmers may be excluded from decision-making in the chain, or they may actively contribute to designing and steering the production processes and forms of cooperation in the value chain. Chain

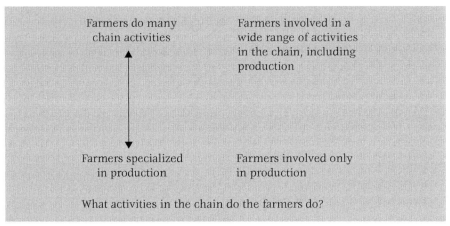

Figure 1. Farmers' involvement in chain activities.

governance determines the conditions under which chain activities are done; it includes aspects such as definition of grades and standards, targeting of consumers, management of innovation, and so on (Figure 2).

If we combine the two diagrams we get a matrix (Figure 3). Farmers may be located anywhere in this matrix. Here are some examples:

1. Susanna (fictitious name) keeps a herd of goats in arid northern Kenya. Every few months, she sells a few goats to a trader who visits her village. The trader dictates the price he pays, and she has no choice but to accept. We call her a *chain actor*, because she engages only in farming and has no influence over chain governance.
2. Pius grows maize on his small farm in western Ghana. He harvests and dries his grain, then mills it into flour before selling it to a trader who visits his village. We call Pius an *activity integrator* because he has moved from farming into other activities in the chain (drying

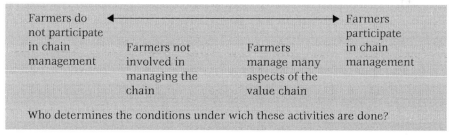

Figure 2. Farmers' involvement in value chain governance.

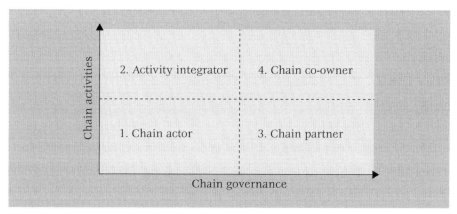

Figure 3. Four forms of farmers' participation in value chains (KIT et al., 2006, adapted from Peppelenbos, 2005).

and milling the maize into flour), yet without exerting any additional influence on the governance of the chain.

3. Mama Bekalo grows pineapples in coastal Tanzania. She sells her fruit to the farmers' association but does not do any processing. The farmer association bulks and grades the fruits as a strategy to get better deals with purchasers. The association has negotiated a contract to supply luxury hotels in Zanzibar. We call Mama Bekalo a *chain partner*, because she specializes in farming and - through the association - exerts influence over management of part of the chain. Chain partners have a long-term partnership with traders, processors or retailers.

4. The Kaffa Forest Coffee Association harvests coffee beans, removes the pulp, dries and roasts the beans, and packages them for export from Ethiopia to Germany. The association has negotiated to supply several importers with high-quality beans, and has created its own brand that is able to attract premium prices on the German market. We call this association a *chain co-owner*, because it has moved downstream, increasing both its chain activities and its influence over chain governance.

To monitor changes in the position of the farmer in the value chain, we can think of the matrix as a continuum. A farmer may start at the bottom left corner of the rectangle. Then he begins grading his product and in so doing moves a little upwards in the rectangle by adding an activity. He also moves a little to the right, because he improves quality management. But he still remains within the chain actor quadrant. If the same farmer later starts processing his product, he may move into the activity integrator quadrant. Or he and his neighbours may organize as a group and negotiate deals with traders, and may start working with the local research institute to test new technologies and new varieties. This would move them into the chain partner quadrant. A combination of these vertical (more activities) and horizontal (more management) advances would push the farmers into the chain co-ownership quadrant.

One danger with a matrix like this is that readers may think that the ideal position for farmers is that of chain co-owner. That is not necessarily true. Many farmers in industrialized countries grow crops under contract for processing companies, and they earn a good living doing so, even though they are 'merely' chain actors. The best chain position for the farmer is situation specific and may change over time. As farmers evolve from chain actors into chain owners, they add 'economic rent' to their business (cf. Kaplinsky and Morris, 2001). But this brings greater risks, investments, and responsibilities, which farmers should be willing and able to bear.

To illustrate how the framework can help to guide farmer organizations in value chain development, we now present a real-life case study from Mali.

4. Shea butter and rural women in Mali

Farmers in the Dioila area of southern Mali traditionally grow cotton for sale on a contract basis to the government's cotton marketing board. However, falling world cotton prices presented farmers with a situation where they could only sell their cotton at a loss. So they began to look for alternative sources of income. An NGO sought ways to support local farmers affected by this uncertainty. It saw shea, a tree growing widely in West Africa, as a potential source of income. Butter made from shea nuts is used in cooking, in creams to protect the skin, and in traditional medicines. The trees are protected for cultural reasons, and therefore are common in fields as well as in the wild.

4.1 Traditional shea chain

Processing shea nut is a complex process. Traditionally, women collect the fruits from wild trees. They put them into large pits until they have time to process them. The fruit rots, leaving the nut inside. They clean off the flesh, crush the nut to remove the kernels, boil the kernels and roast them over a fire to dry. They then crush the kernels to make a paste, which is washed in water to separate fat from the residues. They then filter the fat and boil it to remove the water. The fat cools and solidifies into butter, which they sell in the village markets.

This process results in a low-quality yellowish-brown butter with a pungent smell. This can be consumed locally, but is difficult to sell in urban or international markets. There are no quality standards in trading. The traders determine not only price but also how much to buy, under what terms, and when. They like to buy when the price is low and the women's shea butter stocks are high. The women have no control over any of these conditions. As the women can get only low prices, they keep the best shea butter for their own use and sell the worst quality butter. This results in a vicious cycle of low prices, bad quality, and mutual distrust between the producers and traders. Short-term opportunism prevails over long-term cooperation. Therefore production and processing has stayed the same for centuries. There was no innovation, and no product or marketing improvements.

4.2 Shea chain upgrading

An NGO organized the women into groups to improve the quality of their butter and to start collective sales. 1500 women were organized into 40 community-based groups, which later formed a district union. They received training on various topics such as quality control, improved production techniques, and management. Rather than allowing the fruit to rot in pits, the women now de-pulp the fruit before crushing the nuts, then boil and sun-dry the kernels. This results in an odour-free, clean, white butter that is more appealing to urban consumers.

The groups buy the improved butter from individual group members, then sell it to the union, which in turn sells it to other areas of Mali or exports it other countries in West Africa. By direct sales they cut out traders from the chain. The NGO invested in storage facilities and made information on market prices available to the women, so they could sell their butter when the price was right. Having market information improved the women's bargaining position.

4.3 Achievements and bottlenecks

Quality improvements and the women's stronger skills have enabled the union to more than double the shea butter price from 300 CFA francs (about € 0.45) per kg for traditional butter to 700 francs (€ 1.07) per kg for the improved butter. The union sells an average of 15 tonnes of improved shea butter every year. Even though their individual incomes are still very modest, the women have benefited in other ways; they are now organized, control their production and sales activities, and enjoy recognition in the community and in their families as income earners in addition to being good wives and mothers. There are environmental benefits too: the new procedure (which is based on sun-drying) uses just one third of the firewood compared to the traditional technique.

Despite all the benefits, the chain was unsustainable. Because of the project's short timeframe (4 years), the NGO felt compelled to step into the marketing chain itself. It took control of shea butter marketing, and the determination of sales decisions, price conditions,

Producer organisations and market chains

quality standards, etc. The NGO failed to transfer this capacity to the women's groups; management skills thus stayed within the NGO. The women were passive clients of well-intentioned advice and direction. When the funding ended, the women were unable to take over the activities themselves.

The implicit goal of the project was to make the women chain owners, controlling the whole chain from producer to retail. But what actually happened was forward integration - the women added collective marketing to their business activities, but the women were not empowered to exert any influence over chain governance issues.

4.4 Starting again

The end of project funding jeopardized the many achievements. The women responded by hiring the NGO's commercial advisor as a staff member of the union. They were able to find enough money to pay half his salary, and they obtained funding from a donor to pay the rest. The donor imposed two conditions: (a) the union had to develop a business plan as an independent company, and (b) it had to improve its understanding of the roles and constraints of other actors in the chain, such as traders, transporters, and quality certification agencies.

These conditions compelled the union to develop a new strategy for chain positioning. The union formed a multi-stakeholder team with traders, middlemen, government agencies, transporters, and an exporter to interview all chain actors to identify their roles and constraints. The results were shared in a series of meetings, which allowed the women to address chain management issues, such as quality control of the product at various stages in the chain, information sharing on prices, final use of the products, and emerging market opportunities.

The women managed to negotiate a mixed credit and subsidy package with a financial institution. This will cover all of their investment needs over the five years. It will also cover some of the union's operational costs, but the proportion will fall in steps, from 23% in the first year, down to nothing in year four (4). The women expect their joint earnings to rise from 30 million CFA francs to 58 million (from about € 46,000 to € 89,000) over five years.

One of the unexpected outcomes has been a new business opportunity. A Malian exporter advised the women to sell boiled, sun-dried kernels, rather than shea butter. He said they would be able to sell this easily to chocolate and cosmetics producers in Europe. A European importer was interested in developing a direct partnership with the women's union. The union committed to deliver as many high-quality kernels as possible; trial exports were beginning in 2005. If this business relationship is mutually satisfactory in terms of price, costs, reliability, etc., the European firm may be willing to invest in a processing plant in the Dioila area, as well as in research and development.

4.5 Lessons learnt

The NGO learned that it can play an important role in the initial stages of a market development intervention - for example, in organizing groups of producers and in providing technical training. But it may be better to support the producers' groups directly at later stages of the intervention, especially to pay for key personnel. Now that the women have taken control over the project, they are in a position to develop as a fully-fledged chain partner, consolidating their own businesses, increasing their influence on the management of the chain, and negotiating chain co-ownership with other actors upstream and downstream.

The NGO also learned that value chain development is more than vertical integration - essentially it is about developing farmer capacities to control chain management (Figure 4). Initially the project aimed to empower women to move from being solely chain actors into chain co-owners with control over the entire chain from production right to the consumer (from 1 to 4). However, there was a lack of investment in women's management capabilities, so while they started producing improved shea butter, they had little control over governance of the chain (from 1 to 2). After the new start, the women organized themselves and enhanced their capacities to manage the chain (from 2 to 3).

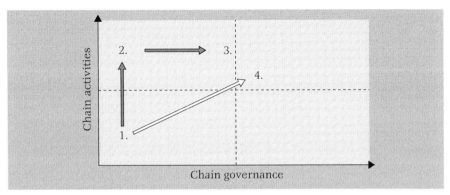

Figure 4. Empowerment of the Malian women in the shea chain.

5. Key capacities for chain development

The case study of shea butter shows that value chain development is a two-dimensional process of promoting vertical integration and enhancing chain governance skills. In this section we define key capacities needed in order for farmer organizations to achieve this.

5.1 Vertical integration

Vertical integration seems to be the preferred strategy of farmers. They like to 'shorten the chain' by cutting out traders or other intermediary agents. They think that adding activities to their businesses will provide them with significant added value and extra income. This, however, is not always true. Adding activities also means adding costs and risks. More importantly, it requires a new set of assets and skills. Four essential capacities are:

1. Technology. Identifying and using appropriate technologies for the value-adding activities (grading, processing, transport, etc.). These technologies must be well maintained and be kept up-to-date. Technological innovation is a permanent concern.
2. Finance. Securing access to: (a) investments in facilities for processing, marketing and distribution, and (b) working capital to run the operations. Reserves must be built up for future investments. Profits must be divided in a rational way between the farmers and the cooperative. Profits should be allocated in accordance to the contribution of each member.
3. Human resources. Building up managerial competence and other necessary human resources to operate these facilities - for example, a specialized marketing manager or quality control staff.
4. Organization: Making sure that the farmer organization has the organizational discipline to get involved in joint value-adding activities. Farmer-members should adhere to quality standards, delivery procedures, and obligations to deliver their produce.

5.2 Chain governance

The return to investments in vertical integration may be disappointing unless due attention is also given to the second dimension of chain development: involvement in chain governance. Five key chain management skills are:

Producer organisations and market chains

127

1. Information management. Often the farmers are in a disadvantaged information position. They have no information about the performance of their own organization, let alone of the market. By contrast, companies downstream in the chain tend to have elaborate information systems. For example, supermarkets register the daily buying behaviours of their customers, while processing companies register the yields, volumes and prices of major crops. The more information someone has, the better he or she can manage a company, and the higher the returns should be. To improve the position of the farmers in the chain, access to and management of information must improve. Some elements of information management are:
   - Record-keeping of labour use and farm inputs. This is necessary to allow a proper understanding of the costs involved, to base farm management decisions upon information, and to help build farmers' ability to negotiate the price of the product.
   - Traceability. This means keeping records to guarantee the buyer as to the source of the product and the inputs that were used.
   - Market information. This involves knowing about prices and trends in the market so that the farmers can better bargain with potential buyers.
2. Quality management. Quality management assures that both the product and the production processes satisfy the consumer. It assures that the farm product can find its way into the market. Quality can be a unique selling-point, through which one group of farmers differentiate themselves from other suppliers. Quality increases the attractiveness of farmers as business partners, hence, their bargaining power. Some aspects of quality are:
   - Grading of the product into homogeneous quality grades, each with a different price for a different market segment.
   - Implementation of quality control systems at critical points in the production system. These make sure that the farmers are on top of the product - that quality is controlled.
   - Implementation of quality certification schemes demanded by the market, such as GAP (Good Agricultural Practices), Food Safety Certification, and EUREP-GAP (quality system of European Union supermarkets).
3. Marketing management. This involves ensuring that the product finds its way into the market. Production processes must be tailored to market demands. There must be knowledge of what consumers

want. Products should be produced, designed and packaged to attract customer preferences.

4. Innovation management. Often innovation is steered from above. New technologies are brought to the farmers by extension officers from contracting companies or the public sector. The farmers are passive recipients of ready-made technological solutions. But the reverse can also happen. Farmers have detailed knowledge of what works best in their fields. They can share these experiences amongst themselves, identify best practices, and start experimenting. They can make study trips to large-scale farmers, research institutes and experimentation centres. In this way, scientific knowledge is combined with practical knowledge. This has the potential to not only boost innovation in the chain, but also make farmers more attractive business partners.

5. Cooperation management. Cooperation with other chain actors is a skill in itself. Often chain relations are marked by distrust. Farmers and traders fight over the price; farmers may cheat the traders by putting low-quality produce at the bottom of the crates, and traders may swindle the farmers by using inappropriate weights and measures. This situation is bad for everyone involved. That is why it is important to seek cooperation along the chain. Some elements of cooperation management are:

    – Chain vision. Chain cooperation starts with recognition that chain actors depend on one another for their business performance. A good chain has synergistic, complementary relationships between specialized chain segments. Chain vision can be developed by taking farmers (or other chain segments) on excursions to companies up and downstream in order to expose them to the reality of other parts of the chain. For example, this will demonstrate to farmers that poor quality (at the beginning of the chain) multiplies into significant losses elsewhere in the chain. A bad tomato that is transported to the city is a waste of money. This loss may lower the price paid for a good tomato. Hence it is more important to get a better price for the good tomato rather than get paid for a poor quality tomato.

    – Trust building. Once there is recognition of mutual dependency between chain segments, then there is scope for a dialogue as to shared interests. Initially the dialogue focuses on trust building, exchanging information and creating shared visions. Later, the

Producer organisations and market chains

129

dialogue may result in joint action plans to improve the chain to everyone's benefit.

– Joint action plans. In dialogue with one another, chain actors can identify ambitions that they may want to undertake together, like the development of a new product, or quality improvements. Or they may identify problems that they want to tackle (e.g., quality degradation during transport). For such problems or ambitions they can draft a joint action plan, in which each of the parties undertakes certain actions.

– Negotiation. In such dialogue all parties can also structure negotiations about transaction conditions (price, quality standards, payment procedures, etc.).

6. Conclusions

In this chapter we proposed a framework for strategizing to empower smallholder farmers in value chains. We conclude that value chain development is pro-poor only when it enhances the capacity of smallholder farmers to co-manage the chain so that they can appropriate a greater share of the returns accumulating along the chain. Achieving this is far from easy. It is a long-term process requiring cooperation between public, private and civil sector organizations (Figure 5). Farmers need to be trained, organized and willing to innovate and take risks. The private sector must be willing to do business with some degree of commitment, allowing farmers to improve their business performance

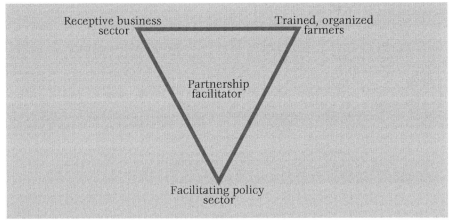

Figure 5. Cooperation for smallholder chain empowerment

by learning from their mistakes. The public sector must be willing to finance training and support. A facilitator must be in place to keep all parties together and to mediate possible conflicts. The producer organization is only one of many stakeholders; one that needs external support and yet needs to be in the drivers' seat. Producer control over their own businesses is a prime condition for chain empowerment.

References

Cochran, W.W., 1979. The development of American agriculture. Minneapolis: University of Minnesota Press.

FAO, 2003. Rise of supermarkets in Africa threatens small farmers. FAO Newsroom. Available from: fao.org/english/newsroom/news/2003/23060-en.html [cited 8 October 2003].

Hoff, C., 2001. Beyond Rosenstein-Rodan: The modern theory of coordination problems in development. Annual World Bank Conference on Development Economics 2000, Washington DC: World Bank.

Kaplinsky, R. and M. Morris, 2001. A handbook for value market chain research. Canada: IDRC.

KIT, Faida and IIRR, 2006. Chain empowerment: supporting African farmers to develop markets. Nairobi: IIRR.

Peppelenbos, L.P.C., 2005. The Chilean miracle: patrimonialism in a modern free-market democracy, PhD dissertation, Wageningen: Wageningen University.

Stockbridge, M., A. Dorward, J. Kydd, J. Morrison and N. Poole, 2003. Farmer organizations for market access: an international review. London: Imperial College Department of Agricultural Sciences.

UNECA, 2005. Economic report on Africa 2005: meeting the challenges of unemployment and poverty in Africa, Addis Ababa: United Nations Economic Commission for Africa.

Fostering co-ownership with producer organisations in international value chains: a strategic answer to the mainstreaming of fair trade fruits?

Dave Boselie

The introduction of a fair trade standard in the fresh fruit sector in 1996 has proven to be a successful initiative to provide smallholder producers with an opportunity to access rapidly expanding supermarket retail segments in many European countries. However, the recent arrival of multinational companies in the fair trade arena threatens to push smallholders once again out of the international retail market. The author argues that co-ownership of the trading firm and the direct influence of producers on value chain development should guarantee the benefits of fair trade for small farmers.

133

1. Introduction

It has been well documented that smallholder producer organisations (POs) in the South face serious threats of exclusion from international value chains (Reardon, 2006, Van der Meer, 2006). Increasing concentration in the retail industry demands economies of scale, while increasing quality requirements demand substantial investments in production system conversion and compliance to quality standards. Smallholders have difficulties in meeting these requirements.

However, the introduction of fair trade standards to the fresh fruit sector in 1996 has proven to be a successful initiative in terms of providing smallholder producers with an opportunity to access the rapidly expanding supermarket segment in many European countries. In addition to favourable conditions for production and trade, the fair trade movement mobilized substantial technical and financial support to allow small producers to build up their capacities. One of the core principles of the fair trade standard - minimum price guarantee - enabled producer organisations to comply with the requirements of good agricultural practices, and respect social criteria like minimum wages and the rights of workers to organise themselves.

While fair trade originally only attracted the attention of smallholder POs that operated in a niche market, over the past two years big fruit operators like Dole and Capespan have entered the fair trade fruit arena aiming to serve the mainstream market. The launch of fair trade products by mainstream multinational food retail companies (like Coop, Aldi, Lidl, WalMart, and others) has been dismissed by some as the cynical exploitation of ethical consumerism.[58] But the fact that fair trade

---

[58] In June 2006 LIDL launched a fair trade line in Germany named 'Fairglobe'. In cooperation with the national fair trade initiative, Transfair, the discount retailer started to offer roasted coffee, bananas and other food products that are sourced at a guaranteed minimum price in order to support smaller farmers in Latin America. However, Transfair's cooperation with the discount chain has already been criticised by people closely connected to Transfair as Lidl is "renowned for social dumping". Apart from the new fair trade line, Lidl in Germany launched its 'Bioness' organic private label range. (www.planetretail.net Daily News 19 April 2006).

is now attracting such interest underlines the growing importance of ethical concerns to consumers.[59]

In numerical terms, the market may still be classified as niche, but the days of fair trade solely as the domain of marginal producers and alternative trade organisations are long gone. And it is perhaps a testament to the success of those early pioneers and campaigners that mainstream operators such as Cadbury Schweppes, Kraft and Nestlé are now showing interest in this segment of the market. According to a recent report on ethical consumerism, labelled fair trade generated an estimated US$ 100 million in additional producer income in 2004, thanks to an estimated US$ 1 billion in global retail sales of all fair trade goods.[60] This represents a 49% increase over the previous year. Moreover, the increasing involvement of large, mainstream companies is likely to accelerate this growth.

In November 2006, a spokesperson from Kraft Foods, the world's second biggest food company, predicted that within the next decade 60% to 80% of the coffee market will be taken up by products with independent certification to fair trade standards. Cadbury Schweppes' acquisition of the leading fair trade and organic chocolate company Green & Black's has prompted speculation that future merger and acquisition activities will focus on grabbing a share of the ethical market. Large multinationals such as Cadbury can increase their profits and their ethical reputation through the acquisition of a smaller, established, ethical company. The acquisition of the Body Shop by L'Oreal is a non-food example of this trend.

Skeptics argue that large companies such as Nestlé view ethical consumerism as the next marketing buzzword, a trend that is worthy of investment. But what is clear is that fair trade as an idea has moved on considerably since it was first introduced in the Netherlands more than

[59] December 2006 witnessed the world's biggest conversion of this kind: Sainsbury's announced it would be the first UK retailer to convert its entire banana range to fair trade. The retail price of bananas was to remain unchanged. The move provides unprecedented volume to fair trade sales and will make Sainsbury's share of the entire fair trade market larger than all the other major supermarkets in the UK combined. Sainsbury's was already the UK's leading fair trade retailer and accounts for the largest market share of bananas with the fair trade mark. As a result of converting to 100 percent fair trade, Sainsbury's will buy five times as many fair trade bananas from growers. This conversion will create a social premium of £4 million in 2007, which will be returned to the growers and their communities. This is an increase of over £3 m compared to 2006. Sainsbury's CEO Justin King said: "This move to 100% fair trade leads the world, and really sets the standard for global fair trade sourcing."
[60] "Global market review of fair trade and ethical food - forecasts to 2012", see: www.just-food.com

20 years ago. The major driver behind the development of the ethical and fair trade market is information. Consumers in many countries worldwide are now aware of the unfair treatment of developing world producers, and are keen to show their support by purchasing the occasional fair trade product. Consumers want to see companies becoming more ethical, caring and compassionate about the product, the consumer, the world we live in, and the environment.

The arrival of multinational companies to the fair trade arena might again threaten to push smallholder cooperatives out of the international retail market. Lower prices will be the inevitable result when these companies aim increasing demand for fair trade products. It is believed that this will have a negative, long-term impact on the producers. Again the principles of economies of scale will put smaller producers in a disadvantaged position. This raises the ultimate question as to what the unique selling point of smallholder POs will be when the fair trade label on their product is no longer unique.

AgroFair Europe ltd applies a business model that might provide an answer that is relevant for others in the industry. One of the central pillars of the company is the concept of co-ownership in a vertically integrated supply chain. The credentials are embedded in its vision statement of: *A Fair Price, A Fair Share and A Fair Say.* Fifty percent of the shares of the company are in the hands of the international producer cooperative CPAF (Cooperative Producers AgroFair) while the other fifty percent is in the hands of European NGOs and sustainable venture capitalists. Based on its business model it has reached the status of preferred supplier to leading retailer COOP Swiss, which has enabled it to capture 40% of the Swiss banana market with its fair trade certified products. Recently SIWA in Finland also chose AgroFair as a company that explicitly promotes co-ownership by southern producers and has development impact upon communities of smallholder producers and workers. In the past three years AgroFair's turnover has been growing with significantly, reaching 66 million Euro in 2006 with a net profit of more than 1 million Euro.

The following sections present the AgroFair business model and the concept of co-ownership in more detail but also discuss the dilemma's that arise when working with a business model of co-ownership with relatively small producers.

2. The AgroFair business concept of co-ownership

The AgroFair business model aims to link producer organisations in twelve countries in Latin America and Africa directly to supermarket customers in various European countries. AgroFair Europe has its headquarters in Barendrecht, the Netherlands, and operates as an importer, distributor and marketing agent. By nature it is a service providing company that coordinates and facilitates the logistical flow and processing of products without actually owning a fleet of trucks or ripening facilities. The company represents the interests of producers from developing countries in the European market.

AgroFair and its producer members foster co-ownership at various levels of the supply chain. Figure 1 gives an organisational diagram of the ownership model of AgroFair Europe. The holding AgroFair Europe has three subsidiaries (in the UK, Benelux and Italy) which provide logistical and marketing support services. In 2006, a new subsidiary was created in the USA (Boston) and country representatives were appointed in Finland and France.

Since the establishment of the company by Solidaridad in 1996, the model has built upon a number of rationales. First of all, there was an economic motive to vertically integrate from primary production to the level of import and distribution so that POs could capture margins and value added in the value chain. Besides giving the POs better control

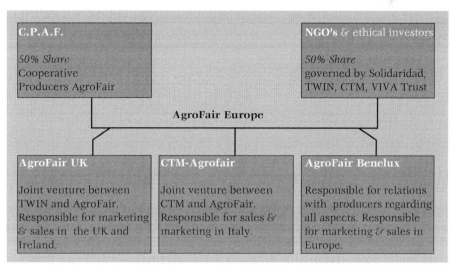

Figure 1. Organisational diagram of AgroFair.

over year-round planning of production and exports, it also gave them access to profits and dividend payments at other levels in the supply chain. By making producers co-owners of the trade and marketing company AgroFair, there was also a higher level of commitment between producers and their first and second level organisations. And last but not least by being a producer-owned company, AgroFair managed to attract the specific attention and dedication of several major European retailers. Some of them (Coop Swiss, Coop UK, Spar, SIWA) follow a cooperative business model themselves.

As stated above the producers collectively own 50% of the shares in the holding company. This grants them income rights to 50% of the annual dividends but it also grants them 50% of the decision-making rights at company board meetings. By maintaining the 50-50 division of decision-making rights between the two groups of shareholders, strategic decisions are always taken by consensus.

Within the international producer cooperative CPAF, voting rights are based on two main principles:
- fifty percent of the voting rights are based on one-man-one-vote principle.
- fifty percent of the voting rights are allocated on the basis of the FOB (freight on board)-value of the traded fruit volumes.

Figure 2 shows the division of voting rights according to traded fruit value in 2005.

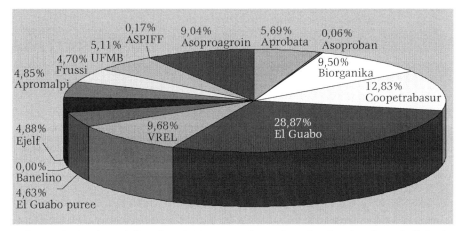

Figure 2. Voting rights for CPAF members (on the basis of traded fruit value).

The producer members of CPAF define their own criteria for membership. The basic criteria include: compliance with the fair trade standard according to the Fair trade Labeling Organisation (FLO), which is the international accredited standard setting and inspection body for the fair trade standard. In addition, the members themselves also prioritize co-ownership, as well as having a fair say in the management of production and trade processes. Many of the current members of CPAF are POs that have cooperative or association-based structures.

With the expansion to new fruit categories new organisational modalities have entered the CPAF cooperative. An interesting example of a new CPAF member is Zebediela Citrus Estate, a citrus plantation in South Africa which is part of a Land Reform and Broad-based Black Economic Empowerment program. The Land Reform Program of the South African government has three main sub programs - Restitution, Redistribution and Tenure. It has as a strategic objective to transform the South African Apartheid land regime and to create an enabling environment for political, social and economic empowerment of Historically Disadvantaged Individuals.[61] Broad-based Black Economic Empowerment (equitable access and participation) in agriculture means economic empowerment for all black people including women, workers, youth, people with disabilities and people living in rural areas through diverse but integrated social or economic strategies, that include, but are not limited to:

a. Increasing the number of black people that manage, own, and control enterprises and productive assets.
b. Facilitating ownership and management of enterprises and productive assets by black communities, workers, cooperatives and other collective enterprises.
c. Human resource and skills development of black people.
d. Achieving equitable representation in all agricultural professions, occupational categories and levels in the workforce.
e. Preferential procurement.
f. Investment in enterprises that are owned or managed by black people.

---

[61] Historically Disadvantaged Individuals (HDI) refers to any person, category of persons or community that is disadvantaged by unfair discrimination before the Constitution of the Republic of South Africa, 1993. Black people is a generic term including Africans, Coloureds and Indians.

Zebediela Estates is one of the largest land restitution farms in South Africa. The farm is approximately 2000 ha but forms part of a total land restitution program of 80 000 ha, including a game reserve and several dams (water reservoirs). The farm was returned, in October 2003, to the original landowners that were forcibly removed from the land by the Apartheid government some 90 years ago. The farm land now belongs to the surrounding community and workers. The operational company Zebediela Citrus Estate is well known all over the world. The farm used to be the largest citrus farm in the Southern Hemisphere, but due to mismanagement and drought the farm went from exporting 2 million cartons in the late 1970's to exporting nothing in 2000. Under serious restructuring and empowerment the farm was able to turn itself around and in 2002 exported 300 000 cartons and in 2003 this number rose to 800 000 cartons. Destination countries for the fruit include: Russia, the Middle-East and Japan. The revival of the operational company was possible only through the involvement of a private investment group, which started to provide management expertise and injected fresh capital. The Boyes Group acquired 51% of the shares in the operational company with a business model that they called the South Africa Farm Management model (SAFM). The community holds 35% of the shares and the workers hold 14% of the shares in the operational company. Over a period of 15 years the management and ownership of the company will gradually be handed over to the workers and the community. Figure 3 presents a schematic overview of the South Africa Farm Management model.

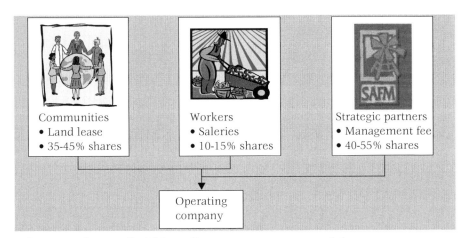

Figure 3. The South African farm management model.

Each of the stakeholders has a complementary role to play: the community contributes land and facilitates the operational company's business expansion, the workers provide their labour and expertise, and the strategic partners of SAFM provide technical knowledge, access to capital, access to markets, and last but not least, management skills.

3. Dilemma's in fostering ownership in the fair trade business model

The previous sections have shown that co-ownership and a joint say in value export chains is one of the pillars of the overall business model of AgroFair. Nevertheless, capacity building among individual producers and their POs to enhance and manage co-ownership has proven to be far from easy. This section presents several of the challenges to fostering ownership in the fair trade business model.

First, the logic of collective action is not always self-evident for individual producers. There is a tension between the self-interest of the individual farmer and the collectively owned producer organisation (cooperative or association). Although a sense of ownership between members and their organisation clearly exists, the short term interest to derive improved personal incomes from the cooperative is often bigger than the interest to make long-term investment in the cooperative. Re-investing profits into the growth of the PO competes with a better price or dividend payments to members. This makes financially weakens many of the primary POs. Based on this reality it is even more difficult to make these weak primary organisations co-owners of a secondary organisation that is shareholder in a European-based import company.

Second, it is difficult to advise, let alone prescribe, specific organisational modalities of co-ownership. Given the vast diversity of business models, such as cooperatives, associations, limited liability companies, there is no single optimal design. Agricultural producers lack the training and capacity to assess the various modalities and decide which is best for their specific situation and circumstances. For the purposes of certification FLO distinguishes only between producer groups or associations and large-scale plantations that operate as private enterprises and mainly work with a hired labour force.

Third, within the fair trade movement but also within CPAF there is intense debate as to what is the ideal producer profile. There is serious concern that FLO policy makers and other stakeholders will make farm size (number of hectares) into a binding criteria for certification eligibility. Instead of using compliance with the social code of conduct (or degree of control and co-ownership) as standard criteria, steps have been made to farm size as the main indicator for deciding whether or not a farm can be certified. A real life example of this criterion is the newly defined fair trade (FLO) standard for Brazil. FLO has limited access to certification for fruit plantations working with hired labour in Brazil using the following criteria for the period from July 1st, 2006 to June 30th, 2008 (see Annex 1): a) limit the farm holding size to 4 fiscal or cadastral entities (*módulos fiscais*) (5 to 100 hectares per modulo depending on the state); b) direct involvement of the owner and/or family members in the management of unit; and c) the owner must live on applying unit or nearby. Setting a maximum farm size conflicts with the original development objective of giving smallholder producers access to export markets: how can we allow them grow if we set maximum farm sizes?

Fourth, AgroFair experience has demonstrated that it is difficult to unite POs of various fruit categories and countries of origin, in one producer-owned business. The diversity of AgroFair members/shareholders has grown tremendously over the past 10 years. The commercial volumes of main products like banana and citrus contrast enormously with small fruit categories like mango, which give the different fruits different political weight within CPAF. The fact that some fruit categories are produced year-round (banana, pineapple), while others are seasonal (e.g. mango a three months season per year) reflects a completely different relationship between farmers and their cooperative. The seasonal character of mango requires sourcing from multiple climatic regions resulting in a partner portfolio of Spanish, French, English and Portuguese speaking POs. AgroFair Europe faces the logistical and organisational dilemma of a multinational company when choosing the language and translations for its shareholder meetings. The minimum cost for a shareholder meeting is 50,000 Euro, mainly spent on travel expenditures to bring producers from various countries to a single location.

Fifth, PO representatives that are a members of the AgroFair family face the challenge of strengthening their capacity as cooperative managers, cooperative board members, and shareholders in AgroFair Europe. Being both a producer and at the same time, directly involved in the management of the cooperative or board member of an operational company can be a serious risk in terms of conflict of interest. It requires a high level of professionalism to separate the various interests one person has to defend. Making a clear division between the tasks and responsibilities of the management and the board of the PO is an important step towards professionalisation.

Sixth, it needs to be emphasized that AgroFair has been aware of these capacity building needs. In 2002 a specific instrument was created to promote organisational and institutional capacity building for POs. Based on a strategic agreement between ICCO and Solidaridad, the AgroFair Assistance & Development Foundation (AFAD) was created. This foundation provides assistance in a number of areas: organisational development, quality assurance and certification, access to finance, and access to markets. A strong emphasis has been on solving the day to day challenges of maintaining or gaining positions in the market. At the same time it has become clear that promoting organisational development takes time! Zebediela's fifteen year time frame for building up a new generation of plantation managers seems to be far more realistic that the one to three year time frame that many donor programs use for obtaining results in PO development.

4. A look towards the future

While the above all sounds extremely positive, many consumers are yet to be convinced, and those that have bought into the fair trade idea appear reluctant to increase the frequency and volume of purchases (especially in markets like the Netherlands). Growth in demand does not keep up with growth in supply. The opposite is happening in the organic market, which is currently experiencing a surge in value. Is there room in the average consumer's shopping basket for both fair trade and organic products? Or, is it still too expensive for the majority of consumers to be ethical, regardless of good intentions? Without doubt, the development of new product categories under the fair trade banner would help to attract new consumers and provide further value to the market. However, sceptics of fair trade argue that there is only so much the movement can achieve, while others believe that future price cuts are inevitable in order to increase demand. If the sceptics are right, this will have a negative, long-term impact on producer income. Other fair trade critics point to the need for a general free trade, rather than fair trade, deal. According to them, reducing the barriers to entry for exporters from the developing to the developed world would make more of an impact on the global food market and provide a positive boost to producers.

Producer organisations and market chains

143

Ethical considerations are starting to dictate food purchases for more consumers, and companies are advised to consider their role within this emerging market. In today's ethically minded society, companies are assessed based on a number of factors such as the origin of ingredients and products, how these products are sourced, environmental impact, and the treatment of the producers/workers. Origin is an important term in today's food and drink business. Organic used to be a niche market but is now entering a more mainstream positioning. Fair trade is in its early stages, but optimistic commentators point to the potential replication of the success experienced so far by the organic movement.

Smallholder POs and alternative trade organisations such as AgroFair should hurry up their process of defining producer profiles in order to distinguish themselves in the mainstream market. With the elements of co-ownership and direct influence in value chain development, AgroFair producers have an excellent opportunity to communicate their own unique selling point. Multinational fruit companies like Dole still source fruit from large scale plantation models; 40% of their trade originates from plantations directly owned and controlled by the company. The other part of their trade comes from large scale plantations owned by private entrepreneurs. Plantation workers and producers do not have involvement and ownership comparable to AgroFair producers. Currently the biggest challenge for AgroFair is to keep the speed of producer capacity building in line with the speed of successful market development. More and more retailers show interest in the uniqueness and origin of the AgroFair model. At the same time the category managers from these retailers will not soften their high demands on quality and consistency of the fruit supplied. Therefore, continuing technical assistance and producer and PO support is needed.

References

Meer, C.L.J. van der, 2006, Exclusion of small-scale farmers from coordinated supply chains. Market failure, policy failure or just economies of scale?. In: R. Ruben, M. Slingerland and H. Nijhoff (eds.), Agro-food chains and networks for development. Springer, 209-217.

Reardon, T., 2006. The rapid rise of supermarkets and the use of private standards in their food procurement systems in developing countries. In: R. Ruben, M. Slingerland and H. Nijhoff (eds.), Agro-food chains and networks for development, Springer, pp. 79-105.

Annex 1. Maximum farm size for FLO in Brazil

State	1 most common módulo (hectares)	4 módulos (hectares) = limit of access
Norte (North)		
Rondônia	60	240
Acre	100	400
Amazonas	100	400
Roraima	80	320
Pará	70	280
Amapá	70	280
Tocantins	80	320
Sul (South)		
Rio Grande do Sul	20	80
Santa Catarina	20	80
Paraná	18	72
Nordeste (Northeast)		
Maranhão	75	300
Piauí	70	280
Ceará	55	220
Rio Grande do Norte	35	140
Paraíba	55	220
Pernambuco	14	56
Alagoas	16	64
Sergipe	70	280
Bahia	65	260
Sudeste (Southeast)		
Minas Gerais	30	120
Espírito Santo	20	80
Rio de Janeiro	10	40
São Paulo	16	64
Centro Oeste (Midwest)		
Mato Grosso do Sul	45	180
Mato Grosso	80	320
Goiás	30	120
Distrito Federal	5	20

African smallholders in organic export projects

Bo van Elzakker

Van Elzakker presents the results of a study done for the Food and Agriculture Organisation of the United Nations (FAO) that reviews four cases of export oriented organic coffee and fruit production. In Sub-Saharan Africa, exporters and smallholders have successfully entered organic and fair trade markets thereby capturing niche market prices. Interestingly, the author does not make a distinction between the way producers are organised in the value chain: whether as a cooperative or through a contracting scheme.

1. Introduction

With the development of organic consumer markets in the North, commercially oriented organic farming projects have developed in sub-Saharan Africa. The majority of these projects are with smallholder farmers. In the African context a smallholder family normally farms 0,5 to 1 hectare of (export) cash crops. In addition, food crops are produced for home consumption, with any excess also being marketed often on a piecemeal basis to buy daily necessities. This chapter presents the main

Bo van Elzakker

outcomes of a study conducted by Agro Eco on the income potential for smallholder farmers of organic agriculture in sub-Saharan Africa. The study was commissioned by the FAO in order to gain insights as to the impact of organic farming on the farmers, particularly the potential for organic farming to increase farmer incomes.

The FAO study is based on a review of four coffee and four fruit production cases. For the coffee cases, two are Arabica and two are Robusta beans. Pineapple is the main fruit for all of the (fruit) cases. Coffee and pineapple are important organic export commodities and they offer opportunities for value addition through improved harvesting and bean processing or fruit drying, as well as for certification. The cases are located in two East African and two West African countries, of which one is Francophone. Upon request of the FAO, information from the eight cases was supplemented by Agro Eco experiences with other organic export projects in Africa.

Although the main focus of the study was on the income effect of producing and exporting organic food products, these cases also provide insight into the organisation of export trade, in particular the role of producer organisations. Export businesses are organised in different ways, including through cooperative unions, through local entrepreneurs, and via multinational trading houses working with contract farmers. The majority of the cases in this study have a hybrid structure where private exporters work with organised (some to a higher degree than others) supplier groups. Therefore, the study offers an opportunity to compare various ways of organizing the link between small scale farmers and quality conscious niche markets.

Through the eight cases, as well as from other experiences, we found that how organic production and export projects were organised was not the main determinant for success or failure. Whether farmers were organised in an official producer organisation, loosely united in a contract farming scheme, or not organised at all, did not really affect the income raising potential of the project.

Data was gathered through the use of a questionnaire, administered by local consultants who already knew the exporters and farmers involved. The exporters were promised anonymity so as to encourage them to discuss their mistakes as well as their successes. All exporters were willing to cooperate; some wanted to be paid for their efforts, especially for organising and providing the financial information. Obtaining useful financial information actually proved to be the most difficult part of the study. For most cases, it took several visits to glean realistic data.

Producer organisations and market chains

148

This chapter is structured as follows. First, it briefly describes the eight cases, how they are organised and the important findings in terms of management and capacity. It then summarises how the different cases deal with issues like organic production, quality control, the role of field agents, and the organisation of value addition and marketing. The final section of the chapter describes the costs and benefits, as well as the potential for increasing smallholder incomes.

For this study the definition of value added is any activity that increases incomes for the rural poor, including: better post harvest management, processing, compliance with buyer's requirements, having a certificate, and individual or collective marketing. It excludes increasing agricultural production.

2. General characterisation of the cases

Table 1 presents several characteristics of the eight cases that were studied. As mentioned above, these cases deal with coffee (K1 to K4) and pineapple (F1 to F4). The third column indicates the structure of the export organisation and its relationship with producers. The type of certification that the product carries is given in the fourth column, and the last column gives the number of farmers involved.

Out of the eight projects, only two were initiated and managed by a producer organisation. One was an old cooperative union (K1). The other consists of three primary cooperatives that broke away from their 'mother' union a few years ago and started to export on their own (K2). It is quite rare for newly established farmer groups to have an independent export position. Farmers are farmers and not processors, exporters or bankers. For non-farm related activities farmers need to employ relatively expensive outsiders whom the farmers' representatives may find difficult to control. In the other six cases, smallholders were organised by entrepreneurs who may or may not promote producer organisations.

The eight cases are considered to be fairly representative of what is happening in the organic coffee and fruit sectors and with similar products like organic cocoa and mango. They are not representative for export crops like cotton, sesame, peanuts or vegetables.

Table 1. Case characteristics.

Case	Product	Structure	Certification	farmers
K 1	Arabica green beans & roasted coffee	Seven primary cooperatives together forming a cooperative union	Fair trade Organic	1058
K 2	Arabica green beans	Three primary cooperatives that split off from the old cooperative union	Organic Fair trade	2125
K 3	Robusta green beans, Spices	Subsidiary of an international coffee house, stimulating primary society formation	Organic Utz Kapeh	922
K 4	Robusta green beans	Subsidiary of international multiple commodity trader, avoiding farmer organisation	Organic Utz Kapeh	4000
F 1	Fresh fruit, air freight	Exporter with contract farmers in three districts, central collection	Organic EUREP-GAP	150
F 2	Fresh & dried fruit, air freight	Producer/processor/exporter with contract farmers in five districts, central collection	Organic	100
F 3	Fresh fruit, herbs, air & sea freight	Producer/exporter with one outgrower group	Organic Fair trade EUREP-GAP	3 own farms and 50 are outgrowers
F 4	Fresh fruit, sea freight, dried fruit	Exporter/packer with 30 groups of contract farmers organised in 4 associations	Organic Fair trade	440

2.1 Organisation

The cooperative union of case K1 is a good example of traditional, large cooperative unions that have problems competing in a liberalised market. They are often overstaffed and therefore work inefficiently, do not work with a business orientation, and have a poor sense of quality and innovation. For these reasons, better functioning primary cooperatives have split off and started marketing their products by themselves. This option became even more attractive because the primary cooperatives already had buyers to help them avoid risky export adventures.

Case K3 is a subsidiary of an international coffee commodity trading house. To make its buying operations more efficient requires forcing farmers to deliver their coffee in groups, thereby promoting producer organisation. K4 is a subsidiary of an international commodity trading house. They do not want farmers to organise for fear of loosing bargaining power and thus they buy from individual farmers at the farm gate, a relatively expensive exercise.

Case F1 is a local entrepreneur who buys fruit from contract farmers. F2 is an Asian-European firm that has its own farm and buys from outgrowers. In addition to exporting fresh produce it has a drying operation. F1 promotes farmers to organise in groups in order to provide extension services more efficiently. Both exporters want their supplying farmers to deliver to central collection points. Case F3 is a farmer-exporter who operates three farms and also works with a group of outgrowers. Because of heavy investments in a packing house, cold storage, planting new fruit trees and drip irrigation, it now finds itself in financial trouble. F4 is a local entrepreneur contracting 30 supply groups that are organised in four fair trade certified associations.

In the case of coffee, fair trade certification is a good way for the PO to increase member incomes. Fair trade requires farmers to be organised in a cooperative. The cooperative receives a guaranteed minimum price for the coffee and a development premium for the group. However, often only part of the cooperative's output is sold as fair trade, usually the better part. The proceeds of these sales are spread out over the entire cooperative membership, thereby reducing the incentive for members to produce quality coffee. Here again, it might be better for the group that produces the right quality for the given market to split off and distribute rewards only among the farmers that produce the quality coffee. In the case of fruit, the impact of fair trade certification on price is not so large, but the premium for group or community development is still important. In case F4 the market demand for fair trade pineapple encouraged the exporter to assist the farmers in organising themselves.

Relatively few traditional cooperatives survive in a liberalised economy. Cooperatives are not known for efficient management nor are they known for being good representatives of their members. To the contrary they are renowned for financial mismanagement and politicisation. Both farmers and private buyers are wary of producer organisations. However, 10-20 years after the demise of the cooperative movement, commercial exporters have started to see the virtue of working with organised supplier groups. These groups are responsible for collecting the product, for the central purchase of inputs, and particular quality activities. However, farmers are still hesitant to organise themselves, due to bad experiences in the past. It is quite difficult to find leaders who do not think of themselves first. As soon as the PO has to handle a large sum of money, trouble starts. Setting up a new producer organisation needs a long term commitment.

On the other hand, when they work well, producer organisations can play an important role in communities. Apart from collecting products from individual farmers, they can provide inputs, channel extension services, credit and savings schemes or road maintenance schemes, and implement small but significant community projects in health care, drinking water or schooling. That is all yet to come.

In most cases these POs are small informal groups of 20 to 25 farmers, where members know one another and learn to build trust. When money is involved, the PO needs a more formal structure, with an executive committee, and often a contact person, which tends to be a lead farmer. In one case there was a first aid/health worker attached due to Utz Kapeh certification.

2.2 Management and capacity building

All cases have a so-called Internal Control System. This is a management and information system that is needed for group certification. It can be used for other purposes as well, like quality improvement. Producing and exporting organic food products requires a number of skills and capabilities to establish and maintain the system. However, there is a limited capacity in the countryside to develop and operate such (management) systems. Well-educated people mostly live in the cities.

The introduction of organic agriculture and to a lesser extent fair trade requires a documentation system, which demands more management at the individual farmer, farmers' group and exporter' levels. The establishment of a documentation system is a difficult process but all cases benefited from this sort of management because it results in a better management of the resources..

One can observe a general lack of accounting and management skills, even among shrewd exporters, and this is worrying. In none of the cases were staff able to produce the costs of operation over the last year without great effort. Exporters, big and small, are not well aware, for example, of logistical costs. Efficiency gains are possible, as can be seen in the different costs for the same activity in very similar projects (see below). Typically, exporters think that they will earn more money by reducing the salaries of the field staff or decreasing the premium paid to the farmers, but such cost reductions have adverse effects. As a result, most cases suffered from good field officers leaving due to poor remuneration and farmers 'side selling' because the organic price was not high enough.

The type of exporting firm (cooperative, international trading house or local entrepreneur) does not seem to make a difference in terms of the success of an organic project. All projects depend on good leadership to get off the ground, but the sustainability of a project very much depends on the leader's ability to delegate and train second-tier managers.

All coffee and two of four fruit projects received support from donor-funded advisory groups. Fair trade groups received far more support (in both financial and human resource terms) than commercial entrepreneurs. Attracting support from different agencies is a business in and of itself. Sometimes support agencies compete to get good projects. It is not uncommon that different agencies support the same cooperative or exporter without knowing about each others' activities.

Managing an organic export project requires new capacities, for example running the Internal Control System. These capacities are not only required from the top person but also from other people in the export organisation. Most interventions focus on building know-how into the exporting organisation, at both head office and field level. The option to contract that capacity building to a commercial service provider, as in K2, is interesting. In countries at a more advanced stage vis à vis organic farming, local advisory capacity is being built up. This can play a key role not only in the development but also the maintenance of organic businesses. When those services are provided for a reasonable price, their presence could be a critical success factor.

2.3 Size of the business

As a rule of thumb, for any organic export business to be sustainable, it needs to export around 250 000 USD of product each year. Where there is a 20 percent organic and quality premium, this allows just about enough room for manoeuvre to pay a premium price to the farmers, maintain a field organisation, pay for the external certification costs, visit foreign markets, and either finance new developments or build up reserves for bad years. The market also demonstrates that buyers are not interested in 50 tons of coffee (except for niches in specialty markets); they are only interested when supplies of 500 tons or more are possible. In the case of fruit, buyers are more interested in 18 ton sea containers each week rather than in 1.5 tons that is sent by air freight. Out of the eight cases, three were of 'sufficient' size and two had the potential to attain it.

3. Revival of farming

The shift from conventional to organic farming was not a big transition for farmers in all except one case (where farmers previously used fertilisers on their pineapples). There were good opportunities for both coffee and fruit farmers to improve their income by improving their farming and (post) harvest practices, generally by investing more labour. Often coffee growing was neglected when coffee prices collapsed. Production at the moment is a fraction of what it used to be. Reviving coffee growing, combining it with organic certification, good (post) harvest practices that result in high quality coffee can once more generate a good income but requires investment in renovation of the farm. Generally, labour is available in abundance but young people are not keen to work in agriculture. Coffee production is left to less enterprising family members. Better farming practices and investing time in farming has the greatest potential to improve incomes.

In the case of pineapple growers, the extra effort is to introduce mulch and better erosion control. Mulching has a fairly direct effect on yields (including drought resilience), while erosion control pays off in the longer run.

To avoid a 3-year conversion period, most projects were started with smallholder farmers who were not using any agrochemicals. Conversion to organic agriculture is much more than a 'do not do anything except get certification' approach. Organic farming must be combined quality improvement, both pre- and post-harvest. This implies taking more care in farming practices, resulting in improved production. Better prices encourage farmers to increase cultivation. If all goes well, a revival in farming is the result.

While the focus in all cases was to get certified and to comply with quality requirements, a side effect in all except one case, was higher yields. The one exception was the pineapple farmers in case F4 who stopped using synthetic fertilisers but did not replace them with organic fertiliser options.

Even more focus on farming could further increase yields. It should be noted that none of the exporters, including cooperatives, showed much interest in supporting suppliers/members to increase production. This is unfortunate because it is the one intervention that has the greatest effect in terms of increasing incomes.

3.1 Field agents

In all cases, government extension services are (almost) fully absent. Actually, basic farming knowledge is often very limited as if there has been no extension at all. In all eight cases extension services were provided by the exporting firm. The latter tends to focus more on issues like certification and quality, and less on the (organic) production.

In different cases different extension models were experimented with. The model based on early joint learning and sharing proved to be the most successful. Extension is expensive. Ideally, one agent should work with no more than 200 farmers (or even fewer in the fruit cases) whereas in the low value coffee cases one agent is often responsible for more than 500 farmers. Parallel side effects to those of introducing Farmer Field Schools, which was in one case the extension approach embraced, can be seen in organic extension, for example farmers starting to take their future into their own hands - becoming decision makers for their own farm and livelihood.

As stated earlier, the exporter's field staff is usually not well paid and both cooperatives and commercial exporters tend to economise on field costs. However, the success of a project is almost directly linked to the enthusiasm of the field agents. Staff turnover is an enormous set back because of the very specific organic knowledge that the field agent must acquire before becoming effective. The Internal Control System is a very good management instrument to systematise extension work.

3.2 Quality

The organic market demands good quality. Consumers are willing to pay more for an organic product, and they expect a quality product. Quality requires attention at the start of a project as African producers are usually not aware of the quality requirements of Northern markets.

There is no organic market for badly fermented coffee, nor is there a market for giant pineapples. Often the buyer is directly involved in quality improvement. Joint quality programmes have a good impact on both producers and buyers who get to know one another, resulting in a better than average product, often resulting in a sort of 'preferred supplier' relationship. In all cases, both farmers and exporters benefited from direct communication with the buyers, and vice versa. All three parties have to learn how to communicate with each other.

Quality pays. In both the coffee and pineapple cases substantial gains were made in areas like harvested-when-ripe and in better post-harvest treatments (pulping-fermenting-drying in the case of coffee, and handling in the case of fruit). This works only when the farmer is rewarded for extra effort by receiving quality based prices. This is still a problem in producer organisations that often have a rule to buy all produce from members, for one set price. The better price given for better quality is generally higher than an organic premium.

3.3 Value addition

In the case of coffee, if farmers do not respond to the call for better harvesting and post-harvest handling, it may be better to take post-harvest away from their control by installing central processing units. This compels farmers (in their own interest) to deliver red cherries only, as well as creating a rural industry by creating seasonal employment for others in the region, and allows the farmer to specialize only in farming.

Given the EU consumer preference for smaller pineapples, only 25% of production can be exported. Therefore it seems a sound idea for fresh fruit exporters to venture into fruit drying. In two cases farmers commented that fruit drying provides a stable income, while fresh exports are prone to sudden fluctuations in demand and therefore, in price. In none of the cases was drying done on a small scale with home-made driers. Fruit dried with home-made driers has never made it to the export market due to poor quality and hygiene. All drying is done centrally in medium sized imported driers. These are based primarily on solar energy but must have another source of energy to guarantee constant hot air flow.

When investments are done, both the POs and exporters run important risks although in the case of POs these are often absorbed by Northern donors. Case F3 featured rapid expansion and major investments and failed. There are technical and managerial challenges when it comes to investment, for which capacity is rarely available in the countryside. Processing facilities tend to be located in larger cities and do not benefit rural communities. When processing facilities are located in the village, there are often problems with electricity and logistics, and it is difficult to attract management keen to live in a village.

In all cases the main improvements were in existing activities. It can indeed be asked why farmers or exporters go into new businesses like processing while knowledge and inputs are available that would double primary production and achieve better prices related to better post-harvest handling. Farmers are farmers; they should become better farmers instead of trying to become processors or exporters. The ideas farmers have about processing and exporting are often not realistic.

3.4 Certification

Prescribing farmers in developing countries to go through a three-year conversion period without providing any subsidies, as are available in Europe or the US, does not seem fair. The introduction of organic, fair trade and other certifications usually needs extra and external assistance.

It is not clear if smallholder farmers benefit from certifications like Utz Kapeh or EUREP-GAP. Utz Kapeh is a coffee certification scheme based on traceability and minimum farming practices (see www. utzkapeh.org). EUREP-GAP is a retailer initiative to guarantee better food safety in fruit and vegetables by insisting on documented Good Agricultural Practices (see www.eurep.org). These schemes result in more obligations and costs for the exporters or smallholder farmers but do not necessarily bring farmers benefits like a better price or better yield. There is some benefit for the exporter in terms of becoming a preferred supplier. Most cooperatives and commercial exporters are not able to introduce and operate these schemes by themselves; they require outside help and people with a university education to manage them. Particularly EUREP-GAP is difficult to comply within a smallholder situation due to documentation requirements.

3.5 Marketing

Both big and small exporters find it difficult to develop organic markets. Such an initiative requires a willingness and ability to communicate with

Producer organisations and market chains

buyers and certification bodies, which is generally weak among African exporters. All types of certification seem to result in markets that are both more direct and more secure. Generally, producers benefit from visits by buyers. Particularly fair trade buyers visit their producers.

It is mentioned more than once in these cases that joint marketing structures should not only be used for export crops, but should also be used to facilitate local sales. In case K3 the farmers earned more money by jointly marketing their maize in the local market than by delivering their coffee for export.

4. Costs and benefits

Table 2 shows the intervention costs per farmer, while Table 3 shows the intervention costs per ton of product, both for the year 2005. The cost per farmer and the cost per ton of product show significant differences between the different projects. Case F3 is not included in these Tables as there were no exports due to near bankruptcy.

It is encouraging to see that the farmer premium is the highest cost, except in case F4 where field operations were most expensive (due to refrigerated transport). The highest premium per coffee farmer

Table 2. Intervention costs per farmer in 2005 (U$).

	K 1	K 2	K 3	K 4	F 1	F 2	F 4
Farmers' premium	38	60	48	16	60	120	120
Field operations	17	8	14	8	43	10	179
External certification	12	5	6	3	56	95	35
Total intervention costs	67	73	68	27	159	225	334

Table 3. Intervention cost per ton of product in 2005 (U$).

	K 1	K 2	K 3	K 4	F 1	F 2	F 4
Farmers' premium	398	64	66	56	107	59	73
Field operations	175	39	64	51	76	5	113
External certification	123	25	26	21	99	46	22
Total intervention costs	696	129	156	127	283	110	207

(excluding the fair trade premium) is paid in K2, which is the most concentrated coffee group, while K4 buys only a small amount from its farmers.

Premiums are higher in the fruit cases. It is evident that exporter F1 (not yet drying fruit) buys less from its farmers than the other two fruit exporters. The cost of intervention per smallholder is much higher for the fruit farmers. These groups are smaller in number when compared with coffee smallholders. That is also reflected in the cost of certification. Within both coffee and fruit projects there are great differences in cost per farmer, despite the fact that farming conditions among coffee growers and among pineapple growers are very comparable. Also the intervention costs per ton of exported product is very variable (Table 3).

Case K 1 has extremely high costs per ton, because it was exporting very little (due to drought) while costs remained. It was paying a very nice premium to its farmers, not realising that it was incurring a loss on the organic operation (see below). The low cost of field operations in F2 shows that the figures provided by the exporter are not correct. It is assumed that the cost of field staff fell under the company's management costs but it was not possible to get more detailed figures from the exporter.

The intervention costs per ton show high variance, both in coffee and fruit cases, while the farming (and living) conditions are fairly comparable. The conclusion is that the structure of the project in combination with ignorance about inefficiencies causes the differences as to the cost of organic intervention per farmer and per ton. It does not seem to matter how or whether the producers are organised. In none of the cases did the exporter (cooperative, local entrepreneur or international trading house) seem to know exactly how it was doing financially. All projects would benefit from better accounting and better cost-benefit analysis.

4.1 Sharing the benefit

Table 4 provides numbers as to the extra income that farmers and exporters earned from the organic project. The figures on the exporters' income were provided by the exporters themselves. Sometimes these figures coincided with figures coming from other sources, sometimes not. It should be realised that these figures are different from year to

Bo van Elzakker

Table 4. Extra organic income and premiums paid (in US$/year).

	K 1	K 2	K 3	K 4	F 1	F 2	F 4
Premium for farmers	39.807	29.525	13.200	33.386	9.067	12.138	51.090
Exporters gain	-18.800	72.750	n.a.	85.460	20.000	32.500	3.900

year, depending on the weather, world demand, and prices. Prices can also vary significantly over the season.

K1 was paying its farmers a very nice premium while incurring a loss. Due to the complicated structure of the cooperative union's accounts they were not aware of this. The information from F4 is fairly unlikely. This exporter, in particular, did not make the impression of being a philanthropist. Healthy profit margins are needed in order to survive doing business in Africa. In the other cases the exporting firm earns more than the farmers. In the case of a cooperative, the exporter's gain is (partly) paid to the farmer members. What happens with the losses in the K1 cooperative is another question altogether.

4.2 Income effects

Table 5 compares the organic premium with the 'normal' annual cash crop income received by the farmer. The annual income was calculated for the same production level but without (or before) any project or exporter intervention. Particularly for the coffee farmer, annual income is not very high. It is better for the fruit growers.

In K1 the drought (low yield) can again be seen. An average income of 250-300 USD per year for a coffee farmer appears to be a fair estimate. This is well below the poverty line of 1 USD per day. The fruit growers

Table 5. Annual income compared with the organic premium (US$/farmer).

	K 1	K 2	K 3	K 4	F 1	F 2	F 4
Annual income	135	290	267	235	850	685	574
Organic premium	38	60	48	16	60	120	120
Income increase	28%	21%	18%	7%	7%	18%	21%

Producer organisations and market chains

normally earn considerably more, three times more a year than the coffee farmers. The size of the fruit farms varies more than the size of the coffee farms, which results in larger variation of income for the fruit farms.

In most cases the conversion to organic farming brings a significant increase in income. The limited income increase in K4 reflects that this exporter buys relatively little per farmer. In case F1 the exporter buys only a small part of the farmers' production as he did not yet have a drier. F2 and F4 do have driers and they are able to buy more fruit from the pineapple growers.

Organic exports results in an average extra income of 40-50 USD per year for the coffee farmers and around 100 USD for pineapple farmers. These figures include the quality premium but exclude the fair trade premium. This is in line with experiences in other projects in Africa. The potential to earn more is much larger for the fruit farmers. However there are far fewer farmers involved in the fruit cases than in the coffee cases.

These comparisons were made using the same production levels. When production increases by 10-20%, which often happens without much effort in organic projects, the change in income is significant. There are also other, non-monetary benefits, like the shift to more sustainable farming practices, but these were not quantified. One may therefore conclude that the monetary contribution to poverty alleviation of organic certification is definitely there, but has its limits. Farmers, especially coffee growers, will not suddenly rise out of poverty due to organic farming.

5. Conclusions

In Sub-Sahara Africa, exporters and smallholders have to some extent done well with entering organic and fair trade schemes thereby capturing niche markets that pay a price premium. The way that producers are organised, in a cooperative or a contracting scheme does not seem to make a big difference.

Introducing organic certification is not enough to obtain and sustain additional income. Wherever the market allows, fair trade certification should be encouraged. Quality improvement is another important money earner. In the case of coffee the greatest increase in income can be achieved by increasing production, increasing farming intensity. Combining these measures will lead to substantially higher incomes.

Producer organisations and market chains

However, the introduction of such a formula requires that substantial resources be developed or brought into the export entity and the farming communities. Of the eight cases, only one operated without donor support.

Change requires both time and an enabling environment, which are often not in place. Usually the 4th or 5th years are crucial and stability is seen only after 7 to 8 years. As donor commitment typically lasts for three years, exporters and farmers are often left to their own devices far too early. Perhaps two of the eight cases can be considered well-established, while the others are still in the process of development.

Small farmer group certification to access socially and environmentally 'just' markets

Rhiannon Pyburn

For smallholder farmers, group certification via Internal Control Systems is a potentially cost effective way to access environmentally and/or socially 'just' markets. This chapter provides an in-depth analysis of several key challenges for smallholder group certification in the global South. These challenges are related to: defining parameters for group certification; documentation; and, developmental considerations. It also discusses viable options for meeting these challenges.

1. Introduction

Certification to social and environmental standards is a means to verify that value integrity is maintained from producer to consumer. However, the requirements and costs of certification can be exceedingly demanding, particularly for small producers in developing countries with less access to knowledge, capacity building and financial resources.

If global food trading relationships are to include small producers in developing countries, then there need to be avenues that support their participation in international markets. Group certification is a mechanism developed in the Latin American organic agriculture sector in the 1980's by small producer cooperatives seeking certification as a group (as opposed to individual certification) in order to lower costs and to stimulate joint learning internally on technical aspects of organic agriculture. This chapter presents research findings as to some of the challenges and strengths of group certification as well as its potential to assess social as well as organic standards. The chapter advocates a critical assessment of group certification in order to strengthen it as a means to both secure market access, and build the capacities of small producers in developing countries.

The chapter begins with an overview of the position of smallholders in alternative markets internationally, followed by a brief background to group certification via Internal Control Systems (ICSs) to orientate the reader. The heart of the chapter is the discussion of three points relevant to smallholder group certification in developing countries: defining parameters; documentation; and, developmental considerations. By examining these three points, issues related to group certification are raised and light is shed on some of the pitfalls as well as potentials of group certification via internal control systems. The chapter concludes with a reiteration of key insights related to the three points and a short discussion as to how group certification can address the main challenges facing smallholders vis à vis market access.

2. Smallholders in global 'just' markets

Internationally, the market share for socially and environmentally certified products is increasing. In 2002 the retail sales of organic products in Europe was estimated to be US$10-11 billion (Martinez and Bañados, 2004:3) with organic imports from developing countries calculated to be worth US$ 500 million (IIED, 1997; Blowfield, 1999; Robins *et al.*, 2000, all cited in Harris *et al.*, 2001:6).

> For tropical products, market shares of labelled products (i.e. organic and fair trade together) are typically one to two percent of the total North American and European markets. This range from 0.8% in the coffee market to two percent for bananas and fresh citrus. Annual growth rates of 20 percent or more in market volume have been

observed for many consecutive years. For some products, such as organic banana, and growth rates of close to 100 percent were reported...sales volumes for fair trade labelled products have been growing at 10 t o 25 percent a year, albeit from a low base (FAO, 2003:iii).

The organic sector is the largest segment of the socially and/or environmentally certified niche in agriculture. The amount of land under organic cultivation is increasing and in 2003 was approaching 23 million hectares (Yussefi and Willer, 2003 cited in FAO, 2003:29). Organic cultivation in developing countries is also increasing:

> ....leading countries with tropical climates are Brazil (275 500 ha certified), Uganda (122 000 ha), and China (around 100 000 ha). The percentage of land cultivated under organic agriculture is highest in Costa Rica at two percent, followed by Uganda (1.4 percent) and Belize (1,3 percent) (Yussefi and Willer, 2003 cited in FAO, 2003:29).

In developing countries the main certified organic products are coffee, banana, cocoa, cane sugar and pineapple (ibid).

Small producers comprise a significant portion of organic farmers; in Chile, for example, 80% of organic farmers are smaller than 10 hectares[62] (Hernandez, 2000 cited in Martinez and Bañados, 2004:5). The International Federation of Organic Agriculture Movements (IFOAM) estimates that smallholders in developing countries produce up to 70% of organic products imported into Europe (2nd International workshop on smallholder certification, Feb. 2002 cited in IFOAM, 2003:1). Smallholders in developing countries are a growing source of organic and certified products. Yet, smallholders are disadvantaged players in global agri-business.

Whereas a significant percentage of food is produced on small holdings, access to international markets is often difficult and cumbersome. The investments required to attain certification may result in smaller suppliers being forced out of business or to supply less profitable markets (Hatanaka *et al.*, 2005:361). Smallholders have been documented as having to face several kinds of challenges in relation to

---

[62] Note that defining smallholders by landholding is not possible as what can be classified as 'small' varies significantly by region, crop and income from farming. IFOAM's detailed and comprehensive definition of smallholder is offered later in this chapter. This Chilean statistic is offered as an example relevant to that country context.

inclusion in the organic sector: production-related; certification-related; and, export-related (Harris *et al.*, 2001:4; Parrott and Elzakker, 2004:26). Each of these will be elaborated briefly in the following paragraphs.

Firstly, production-related constraints include knowledge about organic production, record keeping and perceived inflexibility of EU and other regulations. Resource-poor farmers often do not have access to information about organic production methods and the technical rules and requirements. Record keeping is an essential aspect of organic certification - the inputs and outputs must be well-documented if an operation is to achieve a certificate. Small farmers (in many cases illiterate) find record keeping to be a challenge and in some cases an impossibility. Some dilemmas related to documentation and record keeping are addressed in more detail later in this chapter. The final point in this cluster is that of perceived inflexibility of EU and other regulations. Farmers are compelled to meet standards and regulations created in other parts of the world, which may not be considered reflective of their own conditions.

Secondly, export-related constraints include market knowledge and transportation. Changes in the EU organic certification legislation, particularly EN 45011 requirements, resulted in access barriers for many developing countries (Barrett and Yang, 1999; OECD, 1998; Roberts and Deremer, 1997 all cited in Martinez and Bañados, 2004:2). Regulatory heterogeneity, which is the result of choices made by national governments as well as differences in natural history, resources, infrastructure and social preferences, entails significant costs (ibid). Small producers seeking to participate in export to the North face a regulatory quagmire as the organic sector is still developing, changing and harmonising. Transportation and logistics are also core constraints related to export. The African context for organic production is recognized as a particular challenge (see Parrott and Elzakker, 2004). Parrott and Elzakker (2004) note that poor infrastructure on the continent is a massive challenge; this includes poor quality and badly maintained roads and vehicles, poor communications, lack of refrigeration and supplies, underdeveloped banking and credit systems, sometimes political and economic instability, and other factors leading to logistical problems (2004:26). This chapter addresses some of the challenges related to developmental issues, particularly in the African context.

Thirdly, challenges facing smallholders related to certification are the costs of certification, choice of certifier and the complexity of procedures (Harris *et al.*, 2001:4; Parrott and Elzakker, 2004:26).

Third party certification (TPC) is: 'a form of certification in which the producer's claim of conformity is validated, as part of a third party certification program by a technically or otherwise competent producer or buyer' (Spivak and Brenner, 2001:21-22). TPC is required for organic certification in order to ensure that farmers are compliant with the standards in place. However the costs of certification can be a real barrier for small producers.

> The costs for TPC tend to be the responsibility of suppliers. The annual cost of TPC typically includes three components: (1) audit costs, (2) transportation and field expenses of auditor(s), and (3) costs associated with preparing farms and firms for certification. Costs tend to vary significantly between CBs, as well as by the size of the farming or processing operation and the areas of concern (e.g. food safety vs. organic). However, in general, the cost of TPC is prohibitively expensive for small suppliers, particularly those located in developing countries (Hatanaka et al., 2006:59).

NGOs are often sources of technical and financial support in terms of meeting the requirements of TPC (Hatanaka *et al.*, 2005:362) and supporting smallholder development towards a credible and sound system.

In addition to the obstacle of the cost of TPC, another challenge facing smallholders is the (often broad) choice of certification bodies (CBs) available in any given region. Sometimes there is a difference in the cost of certification between CBs and often they certify to different standards and/or regulations. It is difficult for small farmers to assess which CB and which standards/regulations might be most useful for their needs. Markets available are dependent on these choices. However, choice of CB is just a first step. The procedures for certification are another hurdle in the certification process.

These key constraints facing smallholders - production-related, export-related and certification-related - also apply to farmers participating in group certification schemes. However group certification aims to alleviate each of the three constraints to different degrees and in different ways. The conclusions of this chapter will refer back to these constraints. But before further discussion, some necessary background as to group certification is offered in the next section.

3. Organic group certification - a brief background

In many cases, in order to access markets and coordinate production, farmers organize into cooperatives or associations; this is also true for organic and sustainable agriculture. IFOAM estimated in 2003 that there were approximately 350 grower groups in developing countries involving almost 150,000 smallholders, exporting products to the North (2nd International workshop on smallholder certification, Feb. 2002 cited in IFOAM, 2003:1). An Internal Control System (ICS) is a mechanism that allows smallholder farmers in developing countries to certify produce for western markets through the certification of cooperatives and farmer organisations, therefore allowing small farmers to benefit from the premium prices found in these so-called niche markets.

Organic certification bodies have certified smallholder producer groups since the mid-1980's with principles and basic benchmarks laid down in IFOAM's Accreditation Criteria initially in 1996 (Elzakker and Rieks, 2003:6) in response to practices already happening in the field. According to IFOAM, the rationale behind ICSs for group certification is two-fold: 1) to facilitate smallholder certification i.e. simplify certification and reduce its cost for smallholders through coordinated documentation; and, 2) to implement and maintain a high quality assurance system for organic standards in smallholder production. (IFOAM, 2007).

Internal Control Systems are a means to verify compliance with European organic standards (EU Regulation 2092/91) on the part of small farmers' groups in Latin America, Africa and Asia (Heid, 1999:8). However, defining Internal Control Systems has been an immense undertaking for the organic sector:

> In 2000, there were various reports that showed that many smallholder groups were utterly confused as to what was required from them, especially when they were exporting to different markets and being certified by different certification bodies. The situation of fast growth had resulted in confusion, frustration, high costs, multiple certification and restrained market development (Elzakker and Rieks, 2003:7).

IFOAM, as the international body representing stakeholders in the sector internationally, convened three meetings over three years - February 2001, 2002, 2003 - with certifiers, smallholders, competent authorities, standard setters and others to define and harmonise ICSs and to address some key challenges. The IFOAM-led process set more precise parameters for the implementation of ICSs, inspection and certification, including: defining ICSs, basic elements of ICSs, evaluation protocols, appropriate re-inspection rates, risk assessment tools, amongst others (see Elzakker and Rieks, 2003). The agreed definition of an ICS emerging from that process is the following:

An Internal Control System is a documented quality assurance system that allows the external certification body to delegate the annual inspection of individual group members to an identified body/unit within the certified operator. (As a consequence, the main task of the certification body is to evaluate the proper working of the ICS.) (Elzakker and Rieks, 2003:11)

The IFOAM meetings led to agreed guidelines that set the parameters for smallholder groups, definitions as to who qualifies as a smallholder, and the necessary elements in an ICS. The guidelines define smallholder groups as when:
- the cost of (individual) certification is disproportionately high in relation to the sales value of the product sold
- farm units are mainly managed by family labour
- there is homogeneity of members in terms of: geographical location, production system and size of the holdings
- a common marketing system
- no maximum number of hectares per farmer is set
- minimum size of the group (number of farmers in the group) is not fixed (depends on the situation), but it must be large enough to sustain a viable ICS. A practical guideline is a minimum of 30 to 50 smallholders but
- maximum size of the group is the Group's own concern, depending on their structure, capacity, and communication channels. Size of the group is an element in risk assessment for certification (Elzakker and Rieks, 2003:9).

The IFOAM-convened meetings defined a farmer to be a 'smallholder' and thereby eligible to participate in a group certification scheme when her/his holding meets at least six of the following eight criteria:
- low-tech production system
- based on family labour
- limited capacity of marketing on his/her own

- limited capacity of farm administration
- limited capacity of communication in the language of the certifier
- limited capacity of storage / processing
- the average annual income from the certified product is below approximately 5,000 US$ taken over a number of years (e.g. 5 years).
- would spend over 2% of commodity export value on external inspection if not certified in a group. Also taken over a number of years (e.g. 5years) (ibid).

The different criteria are meant to be calculated as an average: the intention is not that farmers are deemed smallholders some years but not others (ibid). Finally, the IFOAM meetings concluded with the following elements as being essential to an ICS:

- A documented description of the ICS
- A documented management structure
- One person responsible
- An internal regulation (production standard, conversion rules, sanctions etc.)
- Conversion rules i.e. traditional farming/virgin land/known field history
- A contract between the group and the certification body
- Identified internal inspectors
- Training of personnel, internal inspector
- Some form of formal commitment of growers
- Field records, maps
- Annual inspection protocols
- A farm inspection report/form, filled in per farm
- An approval committee that decides to enter the producer on the Growers List
- Use of internal sanctions
- Regularly updated Growers List
- Use of risk assessment to address risks, threats to integrity
- Use of social control/community surveillance (depending on culture)
- Documented post harvest procedures/product flow/quantities (Elzakker and Rieks, 2003:11-12)

These parameters (the definition of smallholders, groups eligible for group certification and the elements required for an ICS) are the basic architecture of group certification via ICSs.

ICSs offer a two-tier form of certification: internal and external. Three key features allow an ICS to provide robust and credible certification: annual internal inspection of all members; external verification of the systems annually with samples of individual farmers; and an active extension provision within the grower group that addresses both technical and organisational issues associated with organic certification. Each of these will be elaborated in the following paragraphs. Additional benefits can also be generated through internal control systems - examples of these will also be mentioned below.

Internal inspection refers to the system put in place by the farmers' group to ensure that every member (some have as many as several thousand members or more) is inspected annually against an organic standard or regulation. Internal (local) inspectors are trained and entrusted to inspect all small farmers in any given ICS to ensure compliance to various regulations (e.g. EU Regulation 2092/91, the Japanese organic regulation - JAS), and/or standards (e.g. IFOAM-accredited standards or other local or international private standards) depending on the markets that the farmer group seeks to access.

External verification refers to the annual visit by a certification body representative to inspect the system that a farmers' group has put in place. An external (e.g. EU-recognised) inspector will focus on ensuring that the ICS mechanism is operating properly as opposed to inspecting each farm separately (Heid, 1999:3). In addition, a sample of farms will be (re-)inspected to verify that what is written in the internal inspection documents reflects the reality on the ground. Through external verification a farmer group gains or loses its organic certification.

Originally ICS were developed by Latin American farmers as a support system for organic smallholders. A control system was attached to this support system to allow for market access via group certification (Bowen and Van Elzakker, 2002:20), but the support or extension element remains an important component. Extension on technical aspects of organic agriculture allows farmers to improve their understanding of the natural systems with which they work. Certification requires both organisational changes and technological upgrades (Hatanaka *et al.*, 2005:361): '...seeking certification and green labels was not simply a social construction of images. It led to, and could only be a result of, organizational change as well as innovation in production methods....' (Jansen, 2004:148). Extension can address the social and organisational systems that comprise organic certification in addition to the technical aspects related to natural systems. Extension often includes organizational issues such as record keeping. Extension and capacity building internally are necessary to ensure farmers are aware of the organic standards to which they must adhere, and how they can meet the standards.

As Grijp *et al.*, 2004 point out, additional benefits can be the outcome of organisation into grower groups for organic certification. Once the ICS is in place and operational, other non-certification-related advantages can emerge:

> These *[secondary]* benefits many include: the transfer of technical advice and support, for example on sustainable production techniques; information on the demands of the international market; improvements for the local community; improved access to credit and reduced costs in the operation of production facilities (Grijp *et al.*, 2004:398, italics: addition Pyburn).

Group certification structures, once in place, have also been used for outreach on HIV-AIDS[63] and community economic development[64] amongst other development-related benefits.

In short, the primary function of group certification via ICSs is to allow market access - inspectabilty is the core. However, intertwined with inspectability are the technical and organisational changes needed in order to create and maintain an inspectable system. Extension and support are required in order to build a resilient, dependable, certifiable system. One the structure (ICS) is in place, a door is opened to other benefits.

---

[63] Personal communication, van Elzakker (April 2004) re: EPOPA project (see: http://www.grolink.se/epopa/)
[64] See SASA Project Thai Audit report 2003: www.isealalliance.org/sasa

4. Group certification for organic and social standards - pitfalls and potentials[65]

The organic group certification mechanism for small producers in developing countries - internal control systems - was the starting point for examining the strengths and weaknesses of current structures, and the associated challenges and opportunities for addressing smallholder needs within the SASA Project[66]. Two cases of organic sector approaches to smallholder group certification provided the basis for the Project's examination of the organic ICS: the Thai rice audit and the Burkina Faso mango audit. These audits examined the structures in place for inspection, capacity building and traceability. The audit teams met with certifiers, farmers, technical extension teams and internal inspectors to gain insights into the workings of these organic models. The SASA project also examined the ICS tool, initially developed for organic group certification, in terms of its viability for social certification purposes. In addition, a further question addressed how the results of social and environmental audits of smallholder production could be used for development purposes beyond certification. Using the organic ICS as a model and considering internal control mechanisms beyond organic (on the Costa Rica certified coffee audit and the Ugandan organic fair trade cotton audit), project discussions explored ways in which small producers are or could be benefiting (in terms of social development) from sustainable agriculture certification.

Three key points from the SASA pilot audits with respect to the organic ICS and its relevance for social certification are elaborated below: defining parameters; documentation; and, developmental considerations. These points are examples chosen to provide some insight as to the challenges and potential for ICSs for certification to both organic and social standards. These points were chosen, from the many raised over the course of the SASA Project, as they are salient issues

[65] Note that parts of this and the following section are based on a report written for the SASA Project by the author of this chapter (Pyburn, 2004)
[66] The empirical basis of this chapter is drawn from cases in four country and production system contexts: Thailand - fair trade organic rice, Burkina Faso - fair trade organic mangoes, Costa Rica - certified coffee, and Uganda - fair trade organic cotton. The case studies were part of the 26-month *Social Accountability in Sustainable Agriculture (SASA)* Project that included four international, private standard-setting organisations that use third-party verification in their systems: FLO - Fairtrade Labelling Organisations International; IFOAM - International Federation of Organic Agriculture Movements; RA - the Sustainable Agriculture Network of the Rainforest Alliance; and, SAI - Social Accountability International.

that shed light on some of the challenges facing group certification via ICSs.

5. Defining parameters of group certification

Recent guidelines agreed within the IFOAM process were discussed above when a background to group certification was provided. Despite this, group certification covers a broad range of farmer groups and standards can vary covering to a greater or lesser extent, social issues, for example. Two issues related to defining the parameters of an ICS that arose in the SASA Project field audits and stakeholder meetings are discussed in this sub-section: the different forms that an ICSs take and related implications; and, defining the social issues that are relevant in the smallholder context.

Internal control system guidelines (i.e. EU regulations or organic certification body standards) are applied universally, yet in stakeholder discussions[67], two distinct forms of organic ICS emerged: endogenous ICSs, and out-grower ICS schemes (see Pyburn, 2003; Pyburn, 2004). The stakeholder meeting led key actors to further articulate what ICSs are and to explicate the different forms that ICSs have taken as well as the importance in making this distinction.

a. *Endogenous ICS:* farmer associations with well-developed and active internal systems that have developed largely through member initiative. These tend to be more established ICSs, often in Latin America.
b. *Out-grower ICS:* group certification driven by economic objectives as opposed to internal support and producer development with the certificate owned by the exporter. Less social cohesion and social control are generally assumed in this kind of a system. These tend to be younger ICSs and are more commonly the case in Asia and Africa.

Each model exhibits different needs and incentives. An ICS of the first type has its own standards and system - it reflects ownership of the

---

[67] The stakeholder meeting on which this discussion is based, was held at the *Biofach Organic Trade Fair* in Nuremburg, Germany in February 2003. For a full account of the stakeholder meeting discussion, refer to Pyburn (2003) or a more condensed version in Pyburn (2004), section 2.2 pages 27-31. The report can be found at: http://www.isealalliance.org/sasa/documents/SASA_Final_ICS.pdf

producer group. The second scenario is characterized by farmers who are suppliers to a buyer where the buyer controls the ICS through the implementation of external guidelines to regulate the supply chain and outsourced farmers. These are two very different systems with different dynamics that need to be understood by certification bodies. For social as well as environmental certification, the methods used and risk analysis done prior to inspection may well need to be tailored in order to reflect the distinctive character (out-grower versus endogenous) of the ICS being assessed.

The characterization of ICSs into two archetypes that emerged from this stakeholder meeting, while important in that it illuminates the differences between two extreme models of ICS, may be too rigid. A more nuanced conceptualization is to consider these models - endogenous versus out-grower - as the extremes reflecting parameters of a continuum. As was raised at the stakeholder meeting, a romanticized perspective is that all ICSs should follow the endogenous model - be 'grassroots-based' (stakeholder statement in Pyburn, 2003). A more challenging, but likely more insightful exercise is instead to recognize the strengths and weaknesses of the kind of ICS under scrutiny. For example, in the out-grower model, the exporter is an expert in export channels and as such is able to find alternate markets and niches within markets more easily. In the endogenous model, the members are farmers - likely lacking broad experience in the international export-import business. The endogenous model, however, tends to have a stronger communication structure internally, which benefits extension, social control and community development initiatives. These variances give an ICS a distinct character, which bears contemplation and reflection in the certification process.

The second issue related to the parameters or limits of ICSs, is that of relevant social issues for consideration in the smallholder context. Group certification was developed for organic standards and that is already a process riddled with challenges. The first question at this juncture is why include social standards in grower group certification? The response is two-fold. Firstly, many smallholder groups either aspire to, or already have, both fair trade and organic certification. Practically, this means that they undergo two separate certifications each year, each demanding time, organisation and different reporting formats. If dual certifications could be streamlined or coordinated, smallholders would benefit, particularly in terms of decreased documentation and decreased costs. However, auditing social issues and environmental issues are two different processes. The question then becomes: what kinds of social issues can be addressed via the ICSs in place?

The International Labour Organisation is a core reference for social standards internationally. Social standards cover a range of issues related to social justice, labour, women's rights, children's rights, worker rights and minimum wages amongst others. The SASA Project addressed the following social issues:

1. Freedom of association and right to collective bargaining.
2. Working hours.
3. Seasonal workers, contracts and undocumented workers.
4. Child labour.
5. Health and safety.
6. Wages/compensation.
7. Discrimination.
8. Basic treatment and disciplinary practices.
9. Forced labour (Lorenzon *et al.*, 2004:1).

However, stakeholders[68] raised concerns as to the applicability and relevance of many social issues in the smallholder context. For example, labour rights issues as applied to smallholder producer context are difficult from a social auditing perspective given that many smallholder producers mainly use family labour and are not structurally dependent on hired labour. Local labour exchange systems within a given community may also be common as was seen in the Thai and Ugandan audit exercises. These particular issues may not lend themselves to the social standards currently used. However, other social and socio-economic issues such as those defined in FLO standards - producer access to markets, fair prices and democratic decision making within the producer group, amongst others - were agreed within SASA, to be relevant to the smallholder producer context. Defining which social standards are applicable to smallholders is essential prior to considering the organic ICS for social certification. However, if social requirements are narrowed down to their essence, then what is directly relevant for the small producer context can be clarified.

---

[68] In the same stakeholder meeting noted in the previous footnote.

6. Documentation

Documentation is a vital component of an ICS, which demands clear and comprehensive record keeping: 'ICS were originally developed to assist smallholders in marketing, record keeping, all kinds of paperwork, communication with the certifier and competent authorities, so actually addressing those issues which a smallholder can not deal with him/herself' (Elzakker and Rieks, 2003:9). However, documentation is burdensome and can be prohibitive. It is an ongoing concern, frustration and necessity related not only to group certification, but auditing in general:

> 'It appears as if much of the control in society is concerned with whether organizational units have the right procedures and produce the right documents, rather than if they are actually doing something differently (Jacobsson, 2000:45).......the need for review thus leads to the production of documents that can be reviewed' (Power, 1997 cited in Jacobsson, 2000:45).

Documentation can be both a (necessary) burden and a (potential) learning device within an ICS (Pyburn, 2004:19-20). ICSs bring more structure to farm management through the documentation of farm inputs/outputs and relevant actions. Record keeping can be regarded as a positive learning outcome of the ICS, wherein farmer understanding and monitoring of on-farm balances improves, the farmers become more organized, they meet regularly and are part of a learning process regarding certification. The Thai audit, in particular found that planning and learning are outcomes even at early stages as production plans compel farmers to think ahead (ibid).

That said, documentation requires literacy. This can be a constraint to smallholders participating in an ICS (see also Parrott and Marsden, 2002:96). In reality, records are often filled in by extension workers on field visits. So, despite some positive outcomes associated with record keeping, the documentation requirements of ICSs tend to demand a lot of extension officers' time. Documentation workload was a repeated concern throughout the SASA Project, for example amongst Thai organic fair trade rice farmers (Pyburn, 2004:19-20), and during the stakeholder meeting in Germany (see Pyburn, 2003). In the Burkina Faso case visual tools were developed to illustrate key points related to certification and standards to farmers and the tools were shared amongst the *groupements* that made up the cooperative via the mango producer cooperative's technical team. These tools permitted local languages to be used in explaining organic/fair trade principles and were useful for non-literate producer understanding (Pyburn, 2004:39).

Social certification would require even more documentation than is currently necessary for meeting organic standards via an ICS. The issues outlined above regarding documentation continue to apply and could well be exacerbated with additional social certification requirements. When adding new requirements to the system, stakeholders agreed that processes rather than checklists should be sought. And where dual certification (e.g. fair trade and organic certifications) are already happening, it is not a question of more documentation, but rather an issue of cutting out overlaps and redundancies in order to streamline and facilitate the certification process for grower groups.

7. ICSs and processes of development - the learning dimension

ICSs are meant to be internal monitoring systems - a quality assurance program - designed internally to satisfy external requirements for certification. An 'internal' system can only work if it is owned by the producer group. This experience of growing from a group of independent farmers to a complete functioning ICS is a process of development (Pyburn, 2004:32). An ICS is not static. It needs to start somewhere and develop with the dynamic of the group. Internalisation of standards however, takes time. Neither the market nor certification bodies allow for extensive growth before compliance is reached. Imposing a pre-determined structure does not fit with a more developmental approach based on progress. A challenge is how to allow for processes of development in certification (Pyburn, 2004:32).

In many cases, an ICS does not initially function in a certification capacity, and can instead be conceptualized as an 'ICS in development', focusing primarily on capacity building. The ICS is, as such, a means to coordinate commercialization and technical training on organic methods and standards. The cooperative puts in place a technical team who is responsible for training/advice and internal inspection of the organic producers. An ICS database can speed-up external certification processes, even before the internal control system as a whole is fully functional. However, investing in an ICS while at the same time paying for 100% external inspection is extremely costly. A key question is how to determine a producer group's readiness for certification via ICS. The reality is that ICSs takes time to develop into a well-functioning system. How can the development required for an ICS to work well, be balanced with the inspection requirements of organic certification?

The SASA Project used the Burkina Faso audit (Pyburn, 2004:34-41) to explore some challenges related to group certification that are particularly relevant to the African context. The Project found that:

- good regional models are lacking as are opportunities for producer exchange on best practices - ICS is a tool without instructions;
- Recognising 'development factors' - essential given the context - sometimes contradicts organic integrity as inspectors need to verify actual situations rather than potential;
- Determination of a producer group's readiness for ICS certification is critical;
- Commercialisation versus social/environmental benefits can be a source of tension.

While raised in the SASA Burkina Faso audit, these challenges may also be applicable to other contexts where ICSs do not have a long history of implementation. IFOAM has addressed some of these concerns by developing training manuals for both setting up an ICS and for ICS evaluation by external inspectors. In terms of capacity building and a developmental/progress-based approach, coordination amongst SASA organizations - particularly between *FairTrade Labeling Organisations International* (FLO) and organic certification bodies - may allow some of these challenges to be addressed. Using FLO's development framework and experience to support organic ICSs as they develop (through progress requirements) is an example of the coordination potential between organic and fair trade (or other social standards) certification bodies and standard-setting organisations. Many principles or requirements are common amongst the four systems (in the SASA Project). In recognising the common features, the efficiency of building on the foundation already in place becomes obvious, though no less a challenge.

8. Conclusions

This chapter began by outlining challenges facing smallholder farmers, including the main constraints vis à vis certification: production-related; export-related; and certification-related. A robust internal control system addresses each of these constraints in a variety of ways. First of all, production-related constraints are addressed through extension, which is a vital element of group certification. Organisation as a cooperative or grower group allows market knowledge and transportation to be centralized. The certificate holder (generally the cooperative or exporter) is empowered by the members to keep abreast of changes in relevant markets as opposed to each farmer trying to grasp this on his/her own. Relationships are built up between the farmer groups and exporters and through the supply chain.

Secondly, in terms of export-related constraints, grower groups generally have an extension function in addition to the internal inspection function. Via the internal extension mechanism, farmers have access to technical support vis à vis organic agriculture as well as support in record keeping and internal management of records. Through group coordination, farmers have more means to meet EU (or other) regulations and potentially a stronger basis for challenging standards or regulations that are inappropriate to tropical or locally specific conditions.

Thirdly, with respect to the certification-related constraints facing smallholders, advocates claim that group certification lowers the cost of certification and buffers farmers from very complex procedures, providing capacity building on what needs to happen in order for the demands of certification to be met.

> The long-terms solution for such countries [*developing countries*] to sustain an international demand for their products lies in structural, strategic and procedural initiatives that build up the trust and confidence of importers/retailers in the quality and safety assurance mechanisms for their produce (Martinez and Poole 2004:229, my italics and addition).

A robust group certification system is one such mechanism for building up the credibility and trust necessary to gain a foothold in global markets and via this route, overcome the market barriers so as to reap the potential for (organic and social) certification to contribute to poverty reduction.

However, an ICS has the potential to be more than a means to access markets. Learning (extension) was the basis of the ICS in Latin America when the system was developed in the 1980's but has taken a back-seat in recent years as the regulatory function was prioritized. In terms of balancing the need for credible, inspectable systems with developmental or progress needs, Ong Kung Wai of the Humus Consultancy in Malaysia rightly stated that what is needed is 'a science for the art' of internal control systems (Pyburn, 2004: 27): valuing and building up the learning side of the ICS in addition to the regulatory aspects. Tensions between extension and internal inspection continue to be an issue, but both are vital to a strong ICS. The learning or developmental dimension of internal control systems can only, in the longer term, strengthen the credibility of the system as a two-tier form of certification that inspires confidence and trust in the markets that smallholders seek to penetrate. The learning and developmental dimension is thus inextricably entwined with inspectability and credibility.

The other issues addressed in this chapter (defining the parameters and documentation) are bound to the central point related to learning and development of the ICS. Endogenously developed ICSs have tended to be stronger in terms of extension relative to out-grower schemes, which tend to be more market-driven. However, the two extremes can learn from one another - both have critical aspects that together would create a stronger whole. Where a particular ICS falls on the continuum - between endogenous and out-grower - needs to match the risk assessment and social issues addressed within the system. The same applies where social standards are included in the organic standard or where separate social certification is sought.

Finally, documentation is a reality facing small producers seeking certification. While documentation cannot be eliminated, the development of simplified forms and the instigation of literacy training within grower groups is vital. Literacy and other social issues are challenges that can be addressed via the ICS structure as the farmer group moves beyond basic technical capacity building to address other organisational and social issues. As such, certification evolves into a poverty reduction tool.

References

Bowen, D. and B. Van Elzakker, 2002. Smallholder group certification - proceedings of the second workshop. February 13 2002, Biofach Organic Trade Fair, Nuremberg, Germany. Commissioned by IFOAM. Written by AgroEco March 2002. pp.24

Brunsson, N. and B. Jacobsson, 2000. A World of Standards. Oxford University Press, Oxford.

Elzakker, B. and G. Rieks, 2003. Smallholder Group Certification - Compilation of Results - Proceedings from three Workshops February 2001, 2002, 2003. Tholey-Theley, March 2003

FAO, 2003. Environmental and Social Standards, Certification and labelling for Cash Crops. FAO Commodities and Trade Technical Paper prepared by Cora Dankers with contributions by Pascal Liu. FAO, Rome 2003. pp.103

Grijp, N. van der, M. Campen Eritja, J. Gupta, L. Brander, X. Fernández Pons, J. de Boer, L. Gradoni, F. Montanari, 2004. "Addressing Controversies in Sustainability Labelling and Certification" in in Eritja (ed). Sustainability Labelling and Certification. Marcial Pons, Ediciones Jurídicas Y Sociales, S.A Barcelona., pp, 387- 403.

Harris, P.J.C., A.W. Browne, H.R. Barrett and K. Cadoret, 2001. Facilitating the inclusion of the resource poor in organic production and trade: opportunities and constraints posed by certification. Report for the Rural Livelihoods Department, DFID, UK, contract number CNTR 00 1301. February 2001 64 pp.

Hatanaka, M., C. Bain and L. Busch, 2006. Differentiated Standardization, Standardized Differentiation: the complexity of the global system. In: Marsden and Murdoch (eds.) Between the Local and the Global - confronting complexity in the contemporary agri-food sector. Research in Rural Sociology and Development - volume 12. Elsevier Jai, Oxford pp. 39-68

Hatanaka, M., C. Bain and L. Busch, 2005. Third Party Certification in the global agrifood system. Food Policy 30:354369

Heid, P., 1999. The Weakest go to the Wall. Ecology and Farming, No. 22.

IFOAM, 2003. IFOAM'S Position on Small Holder Group Certification for Organic Production and Processing. Submission to the European Union and member states 2003-02-03. 5pp. Available at: http://www.ifoam.org/press/positions/Small_holder_group_certification.html last checked February 21 2007

IFOAM, 2007. 'Definitions and Principles of Internal Control Systems (ICSs)'. From IFOAM website: http://www.ifoam.org/about_ifoam/standards/ics/definitionICS.html, last checked February 21 2007.

Jacobsson, B., 2000. Standardization and Expert Knowledge. In: Brunsson and Jacobsson (eds.) A World of Standards. Oxford University Press, Oxford. Pp.40-49.

Jansen, K., 2004. Greening Bananas and institutionalising environmentalism: self-regulation by fruit corporations. In: Jansen and Vellema (eds.) Agribusiness and Society - corporate responses to environmentalism, market opportunities and public regulation. Zed Books, London pp 145-175.

Jansen, K. and S. Vellema (eds.), 2004. Agribusiness and Society - corporate responses to environmentalism, market opportunities and public regulation. Zed Books, London pp 145-175.

Lorenzon, R.P., C. Niel, K. Corbo and S. Courville, 2004. Summary: SASA Recommendations for Consideration on Social Standards, Guidance and Verification Methodologies. August 10 2004. 11pp. Available at: http://www. isealalliance.org/index.

Martinez, M.G. and F. Bañados, 2004. Impact of EU organic product certification legislation on Chile organic exports. Food Policy 29:1-14.

Martinez, G.M. and N. Poole, 2004. The development of private fresh produce safety standards: implications for developing Mediterranean exporting countries. Food Policy 29:229-355.

Parrott, N. and B. van Elzakker, 2003. Organic and like-minded movements in Africa - development and status. International Federation of Organic Agriculture Movements (IFOAM) October 2003, reprint December 2004), Bonn, Germany. pp.130.

Parrott, N. and T. Marsden, 2002. The Real Green Revolution - organic farming and agroecological farming in the South. Greenpeace Environmental Trust, London, UK. pp.129.

Pyburn, R., 2003. SASA Smallholder Workshop Report. Convened at Biofach Organic Trade Fair in Nuremburg, Germany. February 15 2003. pp. 11. unpublished.

Pyburn, R., 2004. SASA Final Report on Internal Control Systems and Management Systems. August 3 2004. ISEAL Alliance, Bonn, Germany. 94 pp. available from http://www.isealalliance.org/sasa/documents/SASA_ Final_ICS.pdf.

Spivak, S.M., and Brenner, C.F. (2001). Standardization essentials: Principles and practice. New York: Marcel Dekker, Inc.

Producer organisations and market chains

Connecting Costa Rican small scale coffee farmers to the main stream European market: an integral chain strategy for sustainable development and improved competitiveness

Myrtille Danse

In Costa Rica coffee cooperatives struggle to meet the quality demands of Western consumer markets. The chapter discusses how external support can help cooperatives meet these demands and how quality standards may trigger continuous improvement of production within cooperatives.

1. Introduction

Small and medium sized farmers in the Costa Rica coffee sector are faced with growing demand from overseas clients to deliver high quality and safely produced goods and services on time, in the correct quantities and at competitive prices. This chapter focuses on experience within the Sustainable Coffee Project (SUSCOF Project) in Costa Rica. The main objective of the project was to help improve the competitive position of the Costa Rican coffee sector within the European market by introducing environmental and social adjustments at the different levels within the supply chain and improve cooperation at chain level.

An important challenge of this project was the creation of a management system to enable different actors within the sector to respond effectively to the continuously changing environment. Continuous improvement is facilitated through an internal management system that helps to measure the performance of processes and whether defined management objectives are being met. This chapter will briefly explain the management systems created and the results obtained.

2. The Costa Rican coffee sector

Since the beginning of the nineteenth century, coffee has been one of the principal sectors driving development in Costa Rica. Nowadays, the sector accounts for approximately 15% of the country's export income, which corresponds to 3% of world trade in *Arabica* coffee. The majority of Costa Rican coffee farms are small to medium sized (maximum 10 to 15 hectares). Current intensive coffee production in Costa Rica is characterised by improved husbandry techniques, such as high density planting, pruning, intensive use of fertilisers and pesticides, and replanting with high yielding, drought and disease resistant varieties. Advanced production and processing methods resulted at the beginning of the 1990s in an average production of 1,610 kg of coffee per hectare, which is considerably higher than the average production in El Salvador (920 kg/ha) or Guatemala and Honduras (690 kg/ha) and has increased since that time. A negative consequence of these intensive production methods is soil pollution and soil erosion due to the use of significant amounts of pesticides and fertilisers.

3. Sustainability issues in the Costa Rican coffee chain

The Costa Rican farmers generally sell their coffee after harvesting the fresh cherries, to small and medium sized private or cooperative coffee mills called 'beneficio' (Blanco Rodriguez, 1999). Over the past few decades the Costa Rican milling process has gone through a number of important changes. After the Second World War, the government expropriated the mills owned by German residents. Due to mismanagement, many of these mills went bankrupt (Blanco Rodriguez, 1999). Nevertheless, high coffee prices in the 1950s made it possible to import innovative technologies allowing coffee mills to integrate both wet and the dry processes.

Currently, mills manage the entire process from coffee cherry to the exportable green coffee bean. To date, Costa Rica has 95 coffee mills, located in five different coffee cultivation areas. The total production capacity of these coffee mills is over 150 million kilograms of green coffee annually.

When designing and building the mills, environmental effects and energetic efficiency were not considered to be important variables. Consequently, the milling process has caused severe environmental problems at the local level, for example, excessive energy, water and firewood consumption, as well as the production of large volumes of organic waste (pulp), and waste water effluents polluted with organic material.

Until the 1970's this model was highly competitive. However, with the market changes in the 1990's this changed. Between 1958 and 1991, international coffee sales were regulated by international treaties which followed the principle of 'finding a reasonable balance between the world supply and demand of coffee, assuring fair coffee prices for the consumer and producer'. However, in the last four treaties such a balance could not be achieved. To the contrary, production was higher than demand, which led to the abolishment of the regulatory system of export quota at the end of the 1980's. As a result, international coffee prices dropped to an all time low of less than 2 USD per kilogram. This price decrease also led to changes on the consumer market, since consumers had greater access to higher quality coffees, which were offered for relatively low prices. As a result of these developments, in the mid-1990's the Costa Rican coffee sector found itself confronted with the following situation:

1. A continuous increase in the area under coffee cultivation
2. Increasing use of agro-chemicals
3. Increasing costs of production
4. Increasing productivity
5. Overproduction of coffee globally
6. Decreased market price, even below cost level
7. Changing consumer preferences

Additionally, the growing awareness of the environmental problems caused by the coffee production - articulated by conservationists and their organizations, amongst others - resulted in stricter local environmental legislation over that same period. In order to prepare the sector for the legal criteria they had to comply with, the Costa Rican coffee sector and governmental bodies agreed in 1992, to a five-year action plan. This five-year plan compelled coffee mills to implement different technical devices that would strongly reduce water consumption in the mills and the emission of waste water into rivers.

Market developments - in combination with government and NGO pressure on natural resource conservation - created awareness within the coffee sector for the need to change course. It stimulated the sector to seek environmentally sustainable solutions, without having an adverse effect on productivity levels and grain quality. This was considered important because only adjusting certain parts of the production process to respond to regulatory requirements, seemed insufficient to meet the needs of the consumer markets, and to differentiate Costa Rica from other coffee selling countries in the region.

A major challenge for producer organisations in Costa Rica is the growing demand from overseas clients to deliver high quality and safely produced goods and services on time, in the correct quantities and at competitive prices. Producers need to demonstrate compliance with a growing number of international standards and technical regulations, which are becoming a prerequisite for entering regional and international markets.

Standards promote trade and commerce by transmitting information in a consistent way, enabling comparisons between products and services. Rapid market liberalisation and the globalisation of trade flows drive the demand for harmonisation and adoption of international standards and related procedures, as does the desire to contain the global consequences of environmental degradation.

Participation by Central American countries in the international standardisation process has been quite low, and the costs in terms of lost trade and investment have only recently been a matter for policy discussion (Wilson and Maizza-Neto, 1999). An explanation for this slow response could be that standards have not played a major role in the raw material and primary commodity sectors, which have dominated the export structure of these countries.

Nevertheless, growing consumer concern regarding food safety (as well as fairness in trade), has resulted in a growing interest within the food industry in major consumer markets such as Europe and the United States, to safeguard and enhance consumer trust in the products offered. This directly affects the many small and medium enterprises

(SME) in countries producing agricultural export commodities, such as the Costa Rican coffee sector. The mainstream coffee trade increasingly is adding sustainability standards to the existing, more straightforward, business transaction requirements of volume, price and quality. This is done with the intention of enhancing performance in terms of sustainability, product quality and risk aversion by ensuring supply continuity (Jansen *et al.*, 2003, Danse and Wolters, 2004). This results in a shift from an entirely deregulated market towards market-led governance through (private) standards and the establishment of long term relationships between buyers and suppliers. These voluntary regulatory systems may not ensure better economic performance, but they facilitate coordination between roasters, traders, and coffee producers (Muradian and Pelupessy, 2005; Danse and Wolters, 2004). To support the Costa Rican coffee sector in developing new market opportunities, the Sustainable Coffee Project (SUSCOF Project) was created. This project aimed to create sustainable production systems throughout the coffee chain taking into account various aspects of sustainability (environmental, social and economic aspects) in such a way as to be flexible enough to respond effectively to present and future requirements.

4. The sustainable coffee project

4.1 SUSCOF strategy

The Sustainable Coffee Project (SUSCOF) was a common endeavour of the Consortium SUSCOF RL, comprised of: six Costa Rican cooperatives representing more then 10 000 coffee farmers; the Dutch-based Institute for Sustainable Commodities (ISCOM); and, the Costa Rican Centre for Technology (CEGESTI). The project (through different consecutive contracts) was financed with funds from the Sustainable Development Agreement between Costa Rica and the Netherlands, as well as the Netherlands Ministry of Environment (VROM), along with considerable in-kind contributions by the cooperatives and individual farmers, and Ahold Coffee Companies.

The project was based on an integrated chain management approach aimed at building continuous improvements in subsequent production processes in the coffee chain. Applying this approach to the Costa Rican coffee sector generated a new perspective. Previously, environmental and social problems were regarded mainly as isolated technical problems

Producer organisations and market chains

189

that had to be improved due to upcoming national or international regulations. However, to improve compliance to market requirements and to be able to differentiate, a change had to be made. Environmental and social issues had to be seen first of all, as a managerial problem. That means accepting responsibility, not only for the installation of certain prescribed inputs but also for the organisation's eventual sustainable performance. To fulfil this responsibility, it was necessary to set priorities and develop verifiable improvement programs. To substantiate this vision and to assess the organisations' environmental strengths and weaknesses, an overview of all relevant environmental effects of coffee cultivation and milling processes was needed, so as to assess the latter's environmental strengths and weaknesses. This led to six Initial Environmental Reviews, which made it possible to quickly recommend how the most pressing environmental (including human health) problems could be addressed.

Observing the common state of cooperatives at that time, one could state that they were ill-equipped to meet the requirements of international environmental standards. Most had little or no experience with management systems, and accounting practices were restricted to formalities required by the government. Managerial experience based on formal procedures was hardly in place, at all. Decreasing market prices, increasing legal requirements and pressure from local and international pressure groups, created awareness amongst cooperative leaders of the need to change.

In order to improve competitive position, the change process had to put in place a gradual performance improvement program, starting with basic workplace improvements, good housekeeping and good agricultural practices. For this to happen, desk and field research revealed that the most effective way to respond was to create management systems for sustainability at different levels within the chain. In this way, a formal system would be developed to enter at the multiple interdependent level to a gradual process of continuous improvement (see Figure 1).

It was assumed that having these systems in place and using them in a prudent and cost-effective way, would be useful in terms of having a tool for learning and change, particularly for organizations that were just beginning to launch a program to continuously improve the quality, environmental and social impact of their products and processes. Additionally, the systems in place should not only have an internal effect but also enable differentiation for the coffee in the international market. For this reason, it was decided to implement

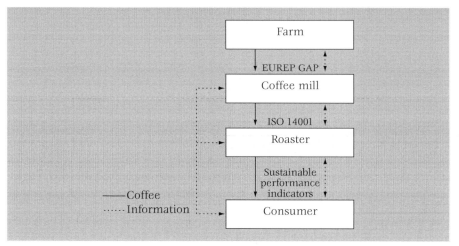

Figure 1. Integral sustainable management approach SUSCOF consortium.

sustainable management systems (SMSs) based on internationally recognized voluntary standards. For this, ISO 14001 based management systems were implemented in each of the six cooperatives and coffee mills; EUREP-GAP was implemented as a pilot project that involved approximately 50 farms per cooperative; a sustainable indicator set was created at system level to enable internal and external planning, monitoring and evaluation; and an indicators set to plan, monitor and evaluate sustainable performance was defined at consortium level.

4.2 Methodological considerations for the multilevel implementation of sustainable management systems

EMS at coffee mill level

Both ISO 14001 and EUREP-GAP define requirements that are easier to apply in bigger, more formalized company structures than in the cooperative structures that were in place within the SUSCOF consortium. To enable efficient and effective compliance with the requirements of the standard, a methodology had to be created that was in line with existing structures and human capacity. In addition, the starting point for working with a group of cooperatives and small scale farmers provided an opportunity for internal group learning and monitoring processes. Due to this, an adapted methodology for the implementation of both standards was developed. For general requirements group training sessions were developed, and visits were made to the different cooperatives and their mills. At the outset, managers were reluctant to share their experiences during group training sessions, since they considered other participants to be competitors. After more than one year, and growing pressure from the coffee crisis, managers became aware that cooperation among local cooperatives would benefit competition with the *real* competitors in their markets - other coffee producing countries such as Guatemala and Colombia. Just prior to certification, in particular, communication amongst managers at operational and strategic levels grew significantly, which helped to create a viable cooperative structure at the consortium level.

The implementation of ISO 14001 requirements began in 1999, and the six mills were successfully certified by 2002. Even in the implementation process, the cooperatives gained better insight into their environmental performance, which resulted in immediate efficiency improvements in production processes causing considerable reduction in water and soil pollution, in particular. Learning networks among the cooperatives emerged, as cooperative managers, mill managers and environmental managers started to exchange knowledge and experience on how to reduce cost and improve environmental performance. These local learning networks also made them less dependent on external business development consultants and thus reduced technical assistance costs.

EMS at farm level

Since the ISO 14001 norms includes requirements focused on environmental aspects of sourcing and purchasing, the implementation of the EMS (at mill level) created a direct linkage between the desired environmental performance improvement at mill level and performance improvement for farmers who had to comply with the mill's new requirements. Since environmental performance requirements were related to farm activities and had to comply with European market requirements, the EUREP-GAP[69] framework was chosen to help farmers to improve their sustainable performance.

At the start of the implementation, EUREP-GAP requirements were used as a base line for farmer cultivation practices. This was done

---

[69] Good Agricultural Practices are formulated by EUREP - a group of European food and retail companies.

based on a survey, the so-called Ecomonitor, which was applied to almost 1000 farmers. The survey allowed insight into the environmental performance of coffee farmers. This information allowed the SUSCOF consortium and project managers to define improvement programs for each cooperative in order to support farmers to improve their environmental performance and to comply in the medium term with the Good Agricultural Practices (GAP) of EUREP. Both the programs that were defined and the monitoring system used, became part of the EMS of the coffee mills. These are incorporated into the annual external audit for the renewal of the ISO 14001 certificate, and this has become a guarantee for cooperative support on environmental issues for the farmers.

Since the SUSCOF consortium represents almost 10 000 coffee farmers, a methodology had to be developed to allow the introduction of EUREP-GAP requirements at group level. This allowed farmers to reach economies of scale with regards to compliance costs, investments in the adaptation of cultivation methods, and technical assistance. In addition, it also resulted in other learning opportunities because training sessions took place in groups and group field visits were organized to the farms of the different cooperatives. In this way additional opportunities were created for farmers to exchange knowledge and experience with one another, but also to catalyse group reflection on new information and knowledge obtained by the external experts conducting training activities.

EMS at distribution and sales level

The management aspects of environmental and social concerns are not only operational in nature but also have significant strategic components. In particular, investment in advanced environmental technology and process adjustments are more rewarding when recognized by the market, for instance through increased sales and/or price premiums or longer term contracts.

In the past most Costa Rican cooperatives were not involved in the commercial activities of the coffee business beyond bulk sales to local intermediaries. As soon as they had processed the coffee, it was sold to middlemen, and they were satisfied with the certainty of immediate cash flows. The growing concern about the origins of coffee by overseas clients over the last decade has made coffee farmers and their organizations more aware of prevailing environmental and social conditions under which they have produced their coffee. A strategic answer to this required a direct link with overseas clients so producers could present the special qualities related to their product and directly respond to their preferences. This idea was included in project design in order to develop tools at sales level to obtain and communicate information and define a planning, monitoring and evaluation process that allowed production of export coffee under a consistent and verifiable sustainability regime.

Originally, the sales aspect was addressed within the framework of discussions with a single big roaster/retailer in the Netherlands; AHOLD coffee company. At a later stage, this changed towards a broader market approach that involved contacting more potential clients in different market segments. To do this in an effective way, a sales organization at consortium level was created, of which each of cooperatives became a member. Since Ahold Coffee Companies was one project partner, they supported the development of the consortium through training SUSCOF's consecutive sales manager at their headquarters in the Netherlands. Also, they facilitated contact with other potentially interested Dutch buyers.

Since the project represented a clear contribution to sustainable development, SMS strategies concerned production, distribution and sales aspects. The SMS activities at mill and farm level resulted in the short term in a differentiated market position, since SUSCOF coffee met the needs of the specific and growing market segment of sustainable coffee. However, if the SUSCOF consortium wanted to maintain its leading position, improvement to sustainable performance at individual cooperative level but also at group level were necessary. The ISO 14001 framework helped to show the consequences of taking such a position and acting accordingly. Nevertheless, the big challenge for the cooperatives participating in the project was not meeting present market needs, but rather to stay on top of future developments and opportunities.

To be able to do this the project helped the consortium to develop an internal monitoring system (IMS) at consortium level. This initiative consisted of designing a set of sustainability indicators that would enable the planning, implementation monitoring and evaluation of the sustainable performance of all cooperatives and farms (Wolters and Danse, 2002). The information collected through the IMS could be used by the consortium for internal planning, and internal and external communication.

5. Results

The project impact with respect to the implementation of a sustainable management strategy should be considered by analyzing the state of the farms, individual cooperatives and the consortium, both before and after project implementation. At the start of the project, base line studies were conducted at both the farm and cooperative levels. This helped to define the implementation strategy, but also facilitated monitoring and evaluation of the impact of the project strategies. In this section, the main results at environmental, social and economic level are presented.

5.1 Social impact

The social impacts of the implementation of the SMS have been:
- Increased managerial capacity level in areas such as organizational planning and administration, both at top management level and for management within the coffee mill.
- Increased awareness amongst personnel as to the existing relationship between their job and the environmental impact as well as tools needed to minimise this impact.
- Decrease in the risk of accidents in different work areas within the coffee mill due to the implementation of an occupational health plan and at the farm level due to training in safe pesticide application and a search for less toxic alternatives.
- Improvement in the working conditions in different work areas due to new protective equipment (safety glasses, belts, gloves).
- Improvements in accident risk control and preparation for environmental emergencies due to the definition and implementation of emergency plans.
- Improved labour conditions due to better defined work schedules, improved reward systems and improved housing conditions for workers living on-site (at the coffee mill or on plantations).
- No children working on the farms or in the coffee mills.

5.2 Environment impact

The environmental impacts of the implementation of the management system have been:

1. Management systems in place that allow control, monitoring, and improvement of the environmental effects of the production processes in coffee mills and on farms
2. Commitment to meet the national environmental law at both cooperative and farm level
3. Introduction of cleaner production technology solutions
4. More efficient use of natural resources
5. Certification of six coffee mills with ISO 14001 (three also in ISO 9002)
6. Certification of 300 farms with EUREP-GAP
7. An average reduction of 11% of fire wood used (max 60%) by coffee mills the first year after certification
8. An average reduction of 11% in kw/ff (max 33%) by the coffee mills the first year after certification
9. An average reduction of 33% of water consumption/ff (max 56%) by the coffee mills the first year after certification
10. Compliance with local legislation of maximum allowed DBO and DQO levels of waste water effluents during the entire harvest period.
11. Controlled waste management at cooperative and farm level
12. Increased shade trees in the plantations.
13. Reduction of pesticides use and replacement of prohibited and highly toxic pesticides by more environmental friendly alternatives.

5.3 Economic impact

With regard to the economic impact, it is important to mention that farmers and cooperative management were not accustomed to quantifying their investments or making adjustments to the processes, nor analyzing related economic data. Before implementation of the project, only quantitative information existed as to fixed costs and some variable costs were calculated per harvest. The financial information was never organized in such a way that enabled management to evaluate (environmental) investments, or to assess the effectiveness of the change implemented.

ISO 14001 and EUREP-GAP norms require the use of records in order to measure the consumption of natural resources and the management of environmental effects generated by farm and milling activities. This information is important for the decision making process in order to improve (sustainable) performance over time. Besides that, this type of data is useful at the consortium level so that the group as a whole can define strategic objectives, define activities to adjust the performance of each cooperative to these objectives and, finally, inform present and future clients and other stakeholders as to the results and performance of the consortium.

Now the most important environmental costs will be presented, related to the wastewater treatment system, the pulp management system, and the management of other waste flows. Each cooperative was requested to present data for the 1998/99, 1999/00 and 2000/01 harvests. However, not all cooperatives had data for the three periods. Therefore, only the data that could be collected are presented below. Financial data are presented in Costa Rican colon. During this period, approximately 350 colon was equivalent to one dollar.

5.4 Investments and operational expenses for environmental improvements

Wastewater and water consumption

The implementation costs for the wastewater treatment systems were an obligatory investment in the mid-1990's for the cooperative to comply with new local regulations described earlier. Costs included design of the system, the purchase of materials, and installation of the system. After the installation, the cooperative had to pay maintenance costs every year. Before the start of the project, a significant part of these costs were related to corrective maintenance tasks, which were adjustments

Table 1. Implementation and maintenance costs in ¢ of the wastewater treatment system for each cooperative in the SUSCOF Consortium, 1998-2001.

Cooperative	98-99	99-00	00-01
Cooperative A	No data	¢2,535,150	No Data
Cooperative B	¢887,500	¢460,000	¢190,000
Cooperative C	¢1,200,762	¢3,334,326	¢7,000,000
Cooperative D	¢34,484,545.88	¢20,438,166.45	No data
Cooperative E	¢163,717,000	¢21,923,000	0
Cooperative F	¢57,580,051	¢5,144,585.65	¢224,111.93

to the functioning of the system since wastewater parameters did not comply with the ones defined by the new regulation.

Due to the implementation of the EMS, the cooperatives learned to plan their harvest and identify improvement opportunities, based on which they conduct preventive maintenance activities. This change in attitude resulted in improved compliance with the legal parameters established. This diminished the risk of getting fined. In addition, it resulted in a lower level of water consumption (see Table 1), which also implies a cost reduction since less electricity is then required to transport water from the source to the coffee mill.

The reduction in water consumption for the consortium between the harvest at the start of the project and the final harvest of the project was 114 776 litres. However, from an economic point of view it is not possible to calculate the savings obtained since the coffee mills don't pay a fee for water consumed, and the cost for electricity is not measured for each activity separately. Over the course of the project, it was recommended that the consortium install a system to monitor these costs. In so doing, it enabled them to identify the areas that require more energy and it facilitated measuring the effects of an investment, for example when making process adjustments or renewing equipment. This data motivates personnel to improve their performance since their work generates cost reductions, allowing the income to be used for other purposes. In addition, it has helped to convince members of the Board of Directors of the cooperatives as to the importance of making certain investments.

Table 2. Water consumption/ff in the coffee mills of the SUSCOF Consortium, harvests 98/99 and 00/01.

Cooperative	97/98	00/01	Variation in %	Total decrease in litres/year
Cooperative A	0.42	0.17	-59	2,950
Cooperative B	0.53	0.48	-9	904
Cooperative C	1.2	0.7	-42	25,455
Cooperative D	1.2	0.5	-58	28,510
Cooperative E	0.41	0.79	92	-27,267
Cooperative F	1.2	0.5	-58	84,224

Electricity consumption

The production processes for coffee mills consume a large amount of electricity. Drying consumes the most since it is a process of a relatively long duration (sometimes it takes more than one day. In addition, equipment at the mills is relatively old since management is not accustomed to frequently update it due to limited budgets and the custom of running the business based on short term planning. Instead, they perform intensive maintenance activities during the periods in between harvests, and in that way they are able to extend the life of the equipment as much as possible. In addition, the coffee mills are designed to be able to receive coffee during the peak harvest, which normally lasts no more than two weeks. This design implies that mills count on an over dimensioned capacity for almost the entire harvest period, which, in any case, is utilized since the machines are positioned in a serial way which requires that all the equipment be turned on in order that the product be processed.

The SUSCOF project managed to obtain various significant changes, which, in the majority of the cooperatives, resulted in a decrease in the kilowatts of electricity per fanega[70] processed (see table 3). First of all, they managed to decrease the amount of water consumed, which implied a reduction in the transportation of water. In addition, the installed capacity was analyzed. Based on this analysis, a preventive maintenance program was designed and implemented. This resulted in more efficient use of the equipment. After this, an electricity efficiency program focused on reducing consumption during the more expensive peak hours of the day, which resulted in cost reduction. In some cases, now they are accustomed to stopping the production process during these hours, in other cases, only equipment that consumes more electricity is turned off.

As can be seen in table 3, five of the six cooperatives were able to decrease their electricity consumption per fanega. Nevertheless, the cost per fanega, instead of diminishing, has actually increased. This result can be explained by the increased petroleum price on the world market. This triggered an increase in the local electricity price. Only cooperatives A and F managed to obtain an economic gain in this field. Cooperative A managed this due to the very small amount of electricity required for the drying process; they dry the majority of their coffee in the sun spreading the product on a platform or using a solar energy

---

[70] 1 fanega is 46 kg of coffee.

Table 3. Electricity consumption (kW/ff) of the SUSCOF Consortium harvests in 98/99 and 00/01.

Cooperative	97/98	00/01	Variation in %	Cost per ff 97-98	Cost per ff 00-01	Total saving/cost
Cooperative A	8.4	7.8	-7	₡313.9	₡247.70	-₡781,160
Cooperative B	7.7	7.13	-2	₡244.2	₡308.2	₡1,157,613
Cooperative C	10.5	9.02	-14	₡280	₡353	₡3,716,430
Cooperative D	9.3	10.97	18	₡158.8	₡337.7	₡7,286,418
Cooperative E	14.1	12.5	-11	₡325	₡366.7	₡2,992,267
Cooperative F	10.5	8.25	-21	₡370 [1]	₡345	-₡3,020,325

(1)Due to the absence of data for the period of 97-98, data for the harvest of 98-99 were used for the calculation.

system, while the other cooperatives tend to dry their coffee using a more industrialized drying process. Cooperative F managed to decrease its electricity consumption due to extra efforts made by management to diminish electricity consumption per day by following a specific monitoring program and taking preventive and corrective measures of consumption per area. This method allowed maintenance of daily consumption below the maximum allowed special electricity rate T6 defined by the local electricity company ICE. Performance within this range enabled cooperatives to obtain a special tariff from ICE.

Firewood consumption

Another resource required in the Costa Rican coffee milling process that demands relatively high investment is firewood. Before the 1990s, the drying process depended 100% on firewood. The potential for using coffee bean husk as an alternative fuel was identified as a result of increasing pressure from a new regulation specifically targeting firewood consumption, and increasing awareness as to existing opportunities to implement cleaner production methods within the sector.

The drying technology currently used by the majority of Costa Rican coffee mills does not allow for 100% firewood substitution by husk since the ovens used cannot reach the same temperature with this much smaller and lighter material. During the project period, the cooperatives were able to install husk dosage equipment and measurement methods for firewood consumption. These changes optimized the balance between the husk and firewood consumption. In addition, a measurement system for firewood consumption per oven was implemented, which allowed monitoring of efficiency for each heat source, and the identification of maintenance needs.

It is important to mention that there is existing technology that allows only husks to be burnt. However, only two of the largest cooperatives in the consortium have this technology installed as the investment required is significant. In addition, cooperative A uses a small solar energy drying system and a drying platform. This allows coffee to be dried throughout the entire harvest using solar energy. Finally, cooperative F has an anaerobic wastewater reactor installed. This reactor captures methane gas that is generated by the wastewater treatment process. This gas is reused in part of the drying process. These more environmentally friendly technological solutions to reduce firewood consumption were installed before the project began. Nevertheless, the project focus helped the cooperatives to optimize efficiency through the implementation of a monitoring system and a training program for personnel.

As table 4 shows, most cooperatives were able to reduce their firewood consumption during the project period, which resulted in a significant reduction in purchasing costs for this resource. Only in the case of cooperative C did a reduction in consumption not result in lowered costs. This can be explained due to a firewood shortage in the zone of influence of the cooperative, which has resulted in an increase in firewood price.

Table 4. Firewood consumption of the SUSCOF Consortium in m^3/ff and ¢/ff, 97/98 and 00/01 harvest.

Cooperative	97/98	00/01	Variation in %	Cost per ff 97-98	Cost per ff 00-01	Total saving/cost
Cooperative A	0.08	0.03	-62	¢46	¢3.8	-¢497,960
Cooperative B	0.071	0.067	-6	¢103.1	¢49.9	-¢962,266
Cooperative C	0.055	0.023	-58	¢80	¢147	¢4,013,970
Cooperative D	0.07	0.03	-57	¢99.3	¢48	-¢2,089,398
Cooperative E	0.11	0.04	-64	¢171.8	¢69.7	-¢7,326,390
Cooperative F	0.035	0.05	43	¢34.57	¢63.70	¢3,50,921

It is important to mention that the majority of the cooperatives obtain their firewood from their members. Often waste wood coming from the coffee plantations is used. In addition, some cooperatives buy waste wood from other companies that cultivate fruit such as mangoes. In the case of cooperative B, sufficient amounts of firewood were not available from either the plantations or other suppliers in the area. Therefore they created reforestation areas that produce wood for the coffee mill. Finally, only cooperative F did not achieve a consumption decrease between the first year of the project and the 00-01 harvest. However, during the 01-02 harvest, they were able to reduce consumption to 0.03 m3/ff and a cost of ¢54/ff.

In addition, the farmers and cooperatives invested in the following activities:
1. Improvement of the pulp treatment system
2. Creation of areas for the temporary deposit of other non-organic waste (plastics, scrap metal, chemical waste among others).
3. Safety equipment and signs marking risk zones
4. Reparation of water leaks and equipment maintenance
5. Certification

In summary, implementation has meant additional costs to allow improvement or replacement of equipment, the installation of monitoring equipment, and certification of the working area. These investments have resulted in a reduction in production costs. In most of the cooperatives costs exceeded profits during the first two years. This was mainly due to the additional expense of putting technology in place that would enable farmers and mill management to comply with requirements in the standards. External audit expenses were also considerable. On average, the mills paid approximately USD 7000 to hire external auditors to conduct external audits and farmers paid approximately USD 1000/ hectare. Largely due to ongoing coffee crises during the project period (2000-2003), the main buyer of SUSCOF coffee financed certification of the first pilot group. In addition, SUSCOF coffee sold for a slightly better price than Costa Rican coffee of comparable quality - namely. 08 USD fanega above the normal price.

6. Conclusion

A lack of insight remains as to what management standards imply and what they actually deliver. It is important to note that voluntary standards such as ISO 14001 and EUREP-GAP focus on processes rather than products. Conformity to these standards certify that a firm has put in place a documented environmental management system and can demonstrate this through repeated audits. While conformity does not guarantee extraordinary positive environmental performance, it can however, provide buyers with greater confidence because they know that a system is in place to observe environmental regulations and that discipline is required to implement and maintain such a system.

The SUSCOF project managed to create a new methodology for the group implementation of ISO 14001 and EUREP-GAP at coffee mill level and farm level respectively. In addition the project created an indicator set that enables planning, monitoring and evaluation of sustainable performance at consortium level.

The introduction of these standards has had both positive and negative effects on farmers and their cooperatives. Farmers and cooperative managers learned to cooperate with other sector representatives and work together towards the improvement of sustainable performance. At the same time, it implied additional work due to the introduction of official working procedures and record keeping systems. Also, technological adjustments had to be made, and in some cases technology had to be replaced. These adjustments resulted in increased efficiencies in terms of labour activities and the use of natural resources, which also resulted in a reduction in costs. The required external audits at farm and cooperative level implied significant extra costs, which could not be recuperated through first year efficiency improvements. However, ongoing improved efficiency of cultivation and processing activities, as well as improved negotiation position at sales level, resulted for most cooperatives in a profitable situation three years after certification.

Besides the short term benefits of the SUSCOF strategy, another question is whether or not the model is viable in the long term. On this topic, the project learned that managers and farmers have to understand the real meaning of the requirements of the voluntary standards, and what the standards can deliver for their own business. Otherwise, they may go through the motions of standards conformance, without bringing about real positive change within the system. A lack of understanding would make design of the required management system a very expensive public relations exercise. Therefore, the implementation of these systems needs to be directly linked with a substantive and verifiable program focused on continuous improvement.

Within the SUSCOF project, a set of sustainability indicators was defined. The objective of these indicators was to provide the SUSCOF consortium with an instrument to create an Internal Management System for the strategic sustainable planning of group initiatives. Managers of the consortium received the proposed set with interest. Nevertheless, the high demand for sustainable coffee in the European market did not create enough incentive for the consortium to overcome constraints related to additional data collection and processing required for the IMS. Therefore the indicators were not implemented, which might cause problems for the competitive position of the SUSCOF consortium in the longer term, since no instruments currently in place allow decisions as to strategic issues at the group level.

References

Blanco Rodriguez, J.M., 1999. Dilemas de la Reconversion del Beneficiado de Café en CentroAmerica, San Jose, Costa Rica: BioMass Users Network, BUN-CA.

Danse, M. and T. Wolters, 2004. Sustainable coffee in the mainstream: the case of the Suscof consortium in Costa Rica, IN: Transforming International product chains into channels of sustainable production: the imperative of sustainable chain management, Greener Management International, Issue 43, September 2004.

Jansen, D., Vellema, S., and Boselie, D., 2003. Linking quality, sustainability and added-value: perspectives for an international coffee index. In: S. Vellema and D. Boselie (eds) Cooperation and competence in global food chains: perspectives on food quality and safety. Maastricht : Shaker: 47 - 64.

Muradian, R. and W, Pelupessy, 2005. Governing the coffee chain: The role of voluntary regulatory systems. World Development 33 (12): 2029-2044

Wilson, S.R. and O. Maizza-Neto, 1999. Enabling enterprise competitiveness in Latin America and the Caribbean through ISO management system standards, Interamerican Development Bank.

Wolters, T and M. Danse, 2002. Towards sustainability indicators for products chains: with special reference to an international coffee chain", IN: Environmental Management Accounting, informational and institutional developments, Volume 9, Springer, The Netherlands.

Building social capital for potato production and marketing: producer organisations' initiatives in north-western Rwanda

Bertus Wennink and Ted Schrader

> *This chapter describes the process of rebuilding the potato supply chain and the role different stakeholders in the Ruhengeri and Gysengi provinces of Rwanda. The authors focus particularly on initiatives taken by producer organisations to develop the supply chain and on the challenges they face.*

1. Introduction[71]

Potato is a long-established food and cash crop in Rwanda. It was first introduced at the beginning of the 19th century and is now being cultivated throughout the country, particularly in the north-western provinces of Ruhengeri and Gysenyi where rainfall and soil conditions

---

[71] This chapter is mainly based on case study results presented in Fané *et al.*, 2004. The chapter is a modified version of the case study presented in Wennink and Heemskerk (2006).

are favourable. Since the transport and infrastructure developments of the mid-1970s, the production and marketing of potatoes has taken on a new importance due to growing urban demand. In 1979 the Government of Rwanda (GoR) initiated a national program to improve potato production (*Programme National d'Amélioration de la Pomme de terre*, PNAP) that concentrated on the development and dissemination of improved varieties. Unfortunately, the civil war in 1994 damaged the country's physical infrastructure and social capital, and the supply of potatoes was seriously affected. Since 1999, both producers' organisations and NGOs have initiated a number of activities to rebuild both the physical and knowledge infrastructure. However, the potato production and marketing chain is still facing serious problems, such as insufficient production and poor quality seed material, lack of storage facilities, and management inefficiencies.

In this chapter we describe the 'rebuilding' of the potato supply chain as well as the role of different stakeholders in Ruhengeri and Gysenyi provinces. We particularly focus on the initiatives taken by producers' organisations to develop the supply chain and on the challenges they face. In the next section we start with a brief description of the policy context and the main features of both farming systems and farmer organisations in north-western Rwanda. At the end of the chapter we draw conclusions from the experiences of producers' organisations in taking supply chain initiatives.

2. Context

Since independence and until the 1980s, Rwandan agriculture experienced significant production increases, mainly through expanding the amount of land under cultivation. But from the 1980s onwards, continuous population growth has led to extreme pressure on the land and a deterioration of natural resource productivity. Coupled with political conflicts, this has resulted in lower yields and a decline in agricultural growth. The 1994 civil war caused the destruction of infrastructure in rural areas. Large sections of the population were displaced, cattle were lost, and planting material and rural infrastructure were all damaged. The new government made the resettlement of refugees, conflict prevention and economic reconstruction its first priorities. In 2004 the GoR adopted a strategy to transform Rwandan agriculture from subsistence-level to market-oriented farming by developing commodity chains and agribusiness, professionalizing agricultural producers, establishing partnerships between the public and private sectors and civil society, and by promoting sustainable production systems (MINAGRI, 2004). These strategic choices were in line with the observed post-conflict dynamism of Rwandan civil society organisations, including producer organisations, which are based on longstanding traditions in rural society.

The Ruhengeri and Gysenyi provinces are situated in north-western Rwanda at the borders with Uganda and the Democratic Republic of Congo. Population density is high, at 500 or more habitants/km^2, and agricultural activities provide 90% of the population with income. Agro-ecological situations are very diverse and include rich soils derived from a chain of volcanoes. Policies for promoting potato production initially targeted smallholdings on volcanic soils where yields (approximately 110 kg/acre, or 11 tons/ha) are twice that of other soils. Potato production also spread to other zones because local urban markets developed over time and smallholders easily adopted crop-production techniques. A survey of potato producing farmer households indicated that up to 85% of farmers rent and own land (< 3 ha). They rely heavily on family labour, and earn income by selling surplus production and renting out labour, investing in renting and buying land. All farmers have a small herd of cattle, some even have a large herd[72]. About half of the potato harvest is marketed, one quarter is consumed by the household, and one quarter is kept for seed.

In Ruhengeri and Gisenyi provinces farmers have a longstanding tradition of organising themselves at the local level into membership-based entities: multi-purpose associations[73], cooperative-type groups called *groupements de base*, and loosely organised *intergroupements* that consist of several associations[74]. At the sector and district levels, single (cooperatives) and multi-purpose (associations) POs have emerged as a result of farmer-led initiatives, which focus on organising collective input supply for crop production as well as marketing. Through these

---

[72] According to IFAD/MINAGRI (2004) the destitute and landless (who are not included in this survey) make up 12% of the Rwandan population. Generally speaking, family farms are poverty stricken and the food poverty level depends on the size of the farm holding and cattle.

[73] Associations are relatively small (10-50 members), their members live in a small area and have fields close to each other. They engage in many activities but have failed to mobilize sufficient capital to ensure the minimum level of functioning of their organization (e.g. travel, participation in district meetings, procurement of collective equipment) and its sustainability (Bingen and Munyankusi, 2002).

[74] Based on overviews presented in Bingen and Munyankusi (2002) and MINAGRI/CTB (2005).

activities, POs are also increasingly involved in organising access to credit facilities for their members. The sustainability of cooperatives largely depends on both the income-generating capacity of production and marketing chains and the management capacity of the cooperatives. Private or state-owned enterprises trading commodities such as tea and pyrethrum have already gained extensive experience with organising producers into associations and *groupements* to manage supply operations within these commodity chains. In the 1990s (after the civil war), associations were reorganised or newly formed to stimulate self-help groups in the post-conflict situation.

Both apex and network organisations have now emerged (at provincial and national levels) such as producers' federations and farmers' syndicates. They often: receive financial and technical support from NGOs and donor agencies; employ technical staff to provide support services to their members; and, support PO capacity strengthening. IMBARAGA(meaning 'force' in the local language Kinyarwanda) is the most important network organisation and operates as a farmers' union (*syndicat*) in both provinces. This *syndicat,* created in 1992, is organised on an individual producer membership basis with branches at sector and district levels. In 2003 IMBARAGA had a registered membership of around 65,000 producers and it employed 23 staff. IMBARAGA initially focused on advocacy and lobbying on behalf of smallholder farmers, but over the years it has also developed member services. IMBARAGA cooperates with two NGOs [the *Bureau d'Appui aux Initiatives Rurales* (BAIR) and the *Forum des Organisations Rurales* (FOR)], which are involved in PO capacity strengthening and manage donor-funded rural development projects.[75]

3. Potato production and marketing: farmer-led initiatives for enhancing chain development

3.1 Seed production and input supply

The national agricultural research institute (*Institut des Sciences Agronomiques du Rwanda*, ISAR) has a long history of experience in breeding high-yield potato varieties that are resistant to pests, and in

---

[75] IMBARAGA, INGABO (another farmer syndicate), BAIR and FOR are members of a national network, the *Réseau des Organisations Paysannes du Rwanda* (ROPARWA). This network facilitates lobbying and advocacy activities and provides support (training and information) for project management.

producing quality breeder seeds. With donor support, the Ruhengeri research station built new greenhouses for improved potato seed production and started on-station and on-farm trials. The national seed service (*Service National des Semences*, SNS) is the next operator in the seed potato supply chain; using improved seed material from ISAR it produces seed potatoes for further multiplication by producers. SNS provides technical support to these producers and supervises certification of the (registered) seed potatoes grown by producers.

In both provinces, registered seed potato producers have been grouped into two cooperative type structures to organise logistics for both collective input supply and for the sale of seed potatoes. The *Coopérative de Développement Agriculture, Elevage et Foresterie* is a private initiative that groups together around 100 producers' associations cultivating collective fields (a total of 60 ha) to grow seed potatoes. The cooperative supplies credit that is reimbursed (i.e., deducted) when the cooperative buys registered seed from the producers. The *Association des Multiplicateurs de Semences Sélectionnées* is a seed producers' association initiated by NGOs and producer organisations. The same NGOs also set up a system for credit supply that is reimbursed by farmers in kind (i.e., in potatoes) rather than in money.

Agricultural inputs, fertilizers/pesticides and equipment for spraying potato seedlings are provided by private enterprises and small traders that are mainly situated in urban centres. Potato producers considered access to input supply in rural areas to be insufficient. In response, NGOs and POs created a rural supply network by building stores in potato producing zones. With support from NGO agents and technicians from the farmers' organisations, farmers manage the stores where farmer groups can buy inputs. Private input traders also created their own organisations to improve access to inputs.

The national research institute ISAR, the national seed services SNS, cooperatives, and private producers increasingly work together to ensure a smooth flow in seed potato production. Still, the quantity of registered seed potatoes produced does not meet demand due to the norms and rules, as prescribed by law, by which seed potato production must abide. It is difficult for smallholders (the large majority of Rwandan farmers) to comply with these rules and thereby increase seed potato production capacity. Seed potato production is still a quasi-exclusive domain of state-owned services such as ISAR and SNS, which forms a barrier for producer participation which could increase seed potato production capacity. An increased transfer of seed potato production to the private sector (producers' groups, individual farmers, etc.) is feasible if existing regulations were adapted and producers' technical skills reinforced. This is one of the reasons (in addition to the expense involved in purchasing registered seed potatoes) that potato producers hold part of their own production for seed supply. This practice affects both production and quality. Producers feel that the norms for seed potato production are not adapted to the conditions of most Rwandan farmers.[76]

Potato producers have mentioned that access and prices limit the use of fertilizers and inputs. This is especially valid for small holdings (< 3 ha), which represent over 75% of all potato producers. Furthermore, the low quality of inputs makes farmers reluctant to buy them. POs, NGOs and private enterprises are all striving to improve these supplies by setting up a network of stores in rural areas to improve access to inputs and to start contracting quality input supply. Grouping orders and other cooperative management practices by POs, as well as quality testing, can create conditions for lower prices and better quality inputs.

3.2 Potato production

Potato production is above all an activity undertaken by smallholder farmers for both household consumption and marketing. In addition to the above-mentioned problems concerning seed potato supply and agricultural inputs, and despite the diversity of farm households and their needs, potato producers also face numerous problems that they share in common relating to production and storage that affect both the yield and quality of potatoes [77]. They feel that the technologies offered by research and extension services are outdated in terms of soil and water conservation measures (as a response to increasing land pressure), potato varieties, organic fertilizer production techniques, and chemical fertilizers. Producers also have little knowledge of potato pests and their treatments, and potato conservation techniques have not been developed at all. In general, it is felt that research and extension do not provide producers with techniques that are effective and affordable under the various agro-ecological and socioeconomic conditions present in the country.

---

[76] For example, producers of registered potato seeds should have at least a 20 ha holding in order to allow a four-year rotation of at least 5 ha.
[77] See survey results presented by Fané *et al.*, 2004.

Since the end of the 1990s NGOs, agricultural extension services and POs have worked together to improve potato production, which is an activity of individual smallholder farmers, in order to improve the income of farm households. Under this joint initiative, IMBARAGA facilitated the creation of potato producers' federations in Gisenyi province (1999) and in Ruhengeri province (2002). The two *Fédérations des producteurs de pomme de terre* that are based on the local farmer groups, *intergroupements*, have representatives at all administrative levels: at sector level (with an elected management committee), at district level (with technical committees that work on issues such as credit supply and production technologies), and at provincial level (where a group of technical committee representatives lobbies for potato producers).[78] However, federations still receive substantial financial and technical assistance from NGOs.

3.3 Research and extension services

The national agricultural research institute (ISAR) has regional research centres in each of the major agro-ecological zones and research stations all over the country. ISAR provides information and develops appropriate technologies. Since 2002, ISAR has been implementing its strategic plan to transform the institute into a client-oriented and user-responsive service provider that plays a key role in agricultural innovation by establishing links with other actors (ISAR, 2002). ISAR is represented in the two provinces through a 'regional innovation centre' at Ruhengeri town, which facilitates consultation between researchers and other stakeholders. At the initiative of IMBARAGA, producers and researchers from ISAR started on-farm trials to study potato varieties and fertilization techniques.

The Ministry of Agriculture's extension service is represented at provincial level by a *Direction* and at district level by an agricultural extension agent (*Responsable du Service Agricole du District*, RSAD). The extension service suffers from a severe lack of human and financial resources. POs and extension services have explored new ways of providing solutions to these problems through lobbying for increased public funding for extension services and common initiatives. On a more extensive scale, the district agricultural extension service and IMBARAGA

---

[78] Rwanda has four layers of political-administrative units: the province (*Intara*), the district (*Uturere*), the sector (*Imirenge*) and the cells (*Utugari*). The district is the most important unit.

technicians developed a train-the-trainer program for Ruhengeri: they trained members of the potato producers' federation, who then trained other member producers. IMBARAGA also developed a farmer-to-farmer extension program. In consultation with local farmer groups, voluntary farmer extension workers are selected on the basis of key criteria such as leadership ability, being proven technology innovators, and having good communication skills. IMBARAGA technicians train these selected farmers, who serve as voluntary extension agents and organise association meetings (both community-based and commodity-based, e.g. potatoes) around specific demonstration plots. The voluntary farm extension agents receive no financial remuneration and plan their own activities.[79]

Although ISAR has decentralized its research as well as its transfer of information and technologies to farmers, this is still weakly organised. Socioeconomic conditions (e.g., land tenure, access to inputs and markets) are the determining factors for innovation. Farmers and extension agents feel that the response from research institutes to real and urgent on-farm problems is too slow and not always well adapted. Decentralizing research management, developing demand-driven approaches for priority setting and planning, putting greater emphasis on adaptive research, and pro-active transfer and adoption assessment of technologies, are all strategic options for ISAR with the potential of strengthening partnerships between POs and research institutes. With their strong grass-root links, POs can also emphasize variation in rural innovation processes.

All key actors involved in research and extension are aware that three different extension approaches currently coexist in the two provinces:

- The agricultural extension service-led approach, which is still inspired by the transfer of technology philosophy (used by RSAD).
- The (still evolving) agricultural research organisation-led approach, which is inspired by farming systems research (used by ISAR).[80]
- The farmer-led approach that is based on voluntary farmer extension workers (applied by IMBARAGA).

This last approach is innovative in that it combines the supply chain development perspective and integrates market norms and standards

---

[79] Experience presented during the Ruhengeri workshop of the 2005 MINAGRI/CTB consultancy mission.
[80] The ISAR Transfer of Technologies Unit became operational in 2004.

with which agricultural products and technologies must comply, with local, community-based networks of which the voluntary farmer extension workers are well-respected members. A first assessment of the farmer-to-farmer extension approach shows that its impact is limited by lack of financial remuneration and weak links with more basic and strategic research for input of (new) technologies. However, the coexistence of several extension approaches is in itself not considered an obstacle to smooth and effective knowledge and information flows. On the contrary, it is the start of a more pluralistic agricultural extension and advisory system that needs strong (but decentralized) coordination in order to clearly articulate the needs of the various production chains and local development stakeholders (input suppliers, producers, transporters, traders, local governments, etc.). However, there is still no specific, PO-led forum where stakeholders can interact to discuss their interests and demands.

3.4 Storage, transport and marketing

Potato marketing is handled entirely by the private sector: small traders who buy directly from potato producers and sell to larger, urban-based, traders. The small traders collect ware potatoes in areas that are difficult to access (steep hills and bad roads) thereby reducing transaction costs for large traders. In order to improve trade flows, *intergroupements* of potato producers have built potato collection and storage facilities and undertake collective sale and transportation. In Ruhengeri province this is done by the *Coopérative d'Exploitation et de Création de Marchés Agricoles* amongst others, and in Gisenyi province by the *Coopérative Ibukwa Muhinzi*. These group actions allow market access for remote potato producers' groups.

Marketing-related activities are thus managed entirely by producers' organisations and traders, and are regulated through delivery contracts between potato producers' organisations and traders. IMBARAGA and the federations introduced delivery contracts, and these contracts considerably improved the efficiency of operations. Still, contracts could be improved by including specifications concerning the delivery time and product quantity/quality. Producers still tend to harvest potatoes prematurely in order to earn some income a little earlier, but this practice negatively affects potato storage quality, which leads to losses for traders.

A national price committee (*Commission de fixation du prix de la pomme de terre*) has been put in place consisting of representatives from state services (including the agricultural extension service), producers, traders and transporters. Every year this committee agrees on a fixed price per kilogram for ware potatoes, regardless of the quality. Survey results show that defining a fixed potato price in the hopes that every producer will receive the same price does not work: prices depend on distances between collection points and markets as well as rural road conditions and in reality are negotiated according to the quality of the potatoes being sold. Quality incentives for producers could enhance chain development. Both traders and producers agree that they should include price differentiation in contracts. Premature potato harvesting highlights problems related to an affordable credit system. Some Rwandan producer organisations manage their own savings and credit programmes, with mixed results. Lobbying for infrastructure development and outsourcing the management of credit supply create better conditions for more equitable access than simply fixing single prices.

4. Concluding remarks

Table 1 presents an overview of the current situation (2004/2005) of the potato production and marketing chain as well as the role and challenges for POs. Experiences in recent years with empowering producers and their organisations in developing this chain allow us to draw the following lessons.

First of all, through emphasizing specific activities and focusing on core business - more or less through a learning-by-doing approach - POs reconsidered their role and functions and started to specialize according to levels of intervention. Both local and district POs focus on economic development (seed and input supply and collecting ware potatoes for marketing) while provincial POs provide support services, sometimes in collaboration with third parties. Local POs play a key role in a country like Rwanda: they are more than any other organisation able to grasp the complex diversity of farmer households and the strategies they use to deal with constraints, to assess their needs, and to provide well-adapted services and advice. Meanwhile, POs at the provincial level, together with the networks in which they participate, lobby for a more enabling policy and institutional environment and, in collaboration

Table 1. Overview of the potato production and marketing chain.

Operations	Actors and their core business	Constraints	Initiatives led by Producers' Organisations	Challenges for Producers' Organisations
Seed potato production	ISAR: breeding and seed potato production SNS: seed production and certification POs: seed production and distribution	Insufficient capacity for seed potato production Unrealistic conditions for seed production (outgrowing)	Cooperatives of seed potato producers (outgrowing) Credit supply for seed producers	Organise demand for seed potatoes and inputs Organise supply of seeds and inputs at the lowest level Renegotiate contracting conditions for seed production
Input supply (fertilizers and pesticides)	POs: grouping members' demand Small traders: assuring supply POs and NGOs: infrastructure for distribution	Lack of quality control for inputs Access to inputs (remoteness of producers)	Rural networks for supply of seeds and inputs	Lobby for strengthening the quality control of inputs
Ware potato production	Individual producers: potato production for household consumption and marketing	Access to seed potato supply and quality inputs Land pressure Insufficient 'up-to-date' technologies Appropriate marketing facilities	Organise producers for seed potato and input supply and marketing Joint on-farm trials with researchers Voluntary farmer extension workers	Grasp the diversity of smallholder farmers and their needs Participate in local policy development and implementation (e.g. enabling environment)
Research and extension services	Producer associations: production Producer federations: training and advice ISAR and RSAD: research and extension services	'Out-of-date' technologies Lack of coordination between training and advisory services	Consultation for innovation with ISAR and RSAD PO/ISAR/RSAD joint activities	Reinforce and sustain farmer-to-farmer extension Lobby for demand-driven research and advisory services

Producer organisations and market chains

Table 1. Continued.

Operations	Actors and their core business	Constraints	Initiatives led by Producers' Organisations	Challenges for Producers' Organisations
Storage, transport and marketing of ware potatoes	POs: product collection by associations POs: price negotiations by federations Small traders: buying	'Imbalanced' contracts between sellers (producers) and buyers (private traders) Fixed and uniform prices for potatoes Weakly adapted credit supply services	National price committee by POs Credit supply by POs	Negotiate adequate contracts Lobby for differential prices (grading and quality of potatoes) Outsource farm credit services

with NGOs, undertake capacity reinforcement activities for lower level organisations.

Secondly, successful chain development stresses the importance of linking with other actors within the chain. However, linkages must go beyond the chain to include providing an enabling environment and adequate support services for producers. Moreover, this demands strong, trustworthy POs which ally together and connect with grassroots member institutions, peer organisations and partner organisations. More precisely, it requires POs to invest in social capital through capacity building for effective relationships at all three levels:

- bonding, or relationships amongst homogeneous producer groups: providing quality services to members, taking into account the mandates given by members, and adequately defending member interests;
- bridging, or relationships with others to access resources and services: transparent financial management and accountability towards partners and members;
- linking, or relationships between different operational levels: communication with producers' organisations operating at other levels, upward representation and downward accountability within larger producer unions and networks. (For more information about social capital in POs, see Heemskerk and Wennink, 2004).

References

Bingen, J. and L. Munyankusi, 2002. Farmer associations, decentralization and development in Rwanda: challenges ahead. Agricultural Policy Synthesis. Rwanda Food Security Research Project MINAGRI. Numéro 3E, April 2002. Available from the website: www.aec.msu.edu/agecon/fs2/rwanda/index. htm.

Fané, I., R. Kribes, Ph. Ndimurwango, V. Nsengiyumva and C. Nzang Oyono, 2004. Les systèmes de production de la pomme de terre au Rwanda. Propositions d'actions de recherche et de développement dans les provinces de Ruhengeri et Gisenyi. ICRA Série de Documents de Travail No. 122, ICRA/ROPARWA/ISAR, Montpellier/Kigali/Butare, France/Rwanda.

Heemskerk, W. and B. Wennink, 2004. Building social capital for agricultural innovation. Experiences with farmer groups in Sub-Saharan Africa. Bulletin 368. Development Policy and Practice, KIT, Amsterdam, the Netherlands.

IFAD/MINAGRI, 2004. Agriculture in Rwanda. Background & Current Situation. Bibliographical document compiled within the framework of the activities of IFAD in collaboration with MINAGRI. Text and design by C. Bidault.

ISAR, 2002. Looking towards 2002: ISAR strategic plan. ISAR, Rubona, Rwanda.

MINAGRI, 2004. Strategic plan for agricultural transformation in Rwanda. Main document. Draft, October 2004. MINAGR, Kigali, Rwanda.

MINAGRI/CTB, 2005. Etude d'identification. Programme d'appui au système national de vulgarisation agricole decentralisé. Etude réalisée par Ted Schrader (KIT Amsterdam), Jean-Marie Byakweli (consultant Kigali) and Jean-Damascène Nyamwasa (PLANEEF Kigali). MINAGRI/CTB, Kigali, Rwanda.

Wennink, B. and W. Heemskerk, 2006. Farmers' organisations and agricultural innovation. Case studies from Benin, Rwanda and Tanzania. Bulletin 374. Development Policy and Practice, KIT, Amsterdam, the Netherlands.

Farmers' organisations in West Africa: emerging stakeholders in cotton sector reforms

Joost Nelen[81]

The chapter paints a picture of cotton farmers' organisations in West Africa. These organisations are becoming accustomed to positioning themselves on different fronts and at different levels and have to deal with a wide range of stakeholders. Three examples are presented to demonstrate how farmers' organisations have, at their own initiative, gradually come to grips with the complex issues inherent in the cotton sector.

1. Introduction: the fickle fate of a symbol for development

Cotton is a classic symbol of development in the dry sub-humid rural areas of West Africa. It is a major source of revenue for several states including Mali, Burkina Faso and Benin, and is a main source of income for more than 20 million people living in a vast region extending from

[81] Thanks to Bertus Wennink (KIT), Giel Ton (LEI/WUR), Nata Traoré, Hans Meenink, Jules Sombie, Cheick Kané and Pascal Babin (SNV). Copyright photo: UNPCB, Burkina Faso.

eastern Senegal to southern Chad. French speaking West Africa is responsible for 4% of world cotton production (approximately 25 million tonnes of cotton-lint per annum) and is second only to the United States in terms of exports. Unlike China or India, which produce both for export and for significant domestic industries, West African countries are 'price takers'.

Until the mid 1990's, cotton sectors in French speaking West Africa were entirely controlled by the state. Parastatal companies generally managed input distribution, the ginning and marketing of cotton, as well as price setting. Other actors were national cotton research organisations and state agricultural extension services. The so-called 'integrated approach' was characteristic of cotton sectors and contributed to a relatively successful performance on the world market. The overall coordination of both chain operations and support services was ensured by one actor, usually the parastatal company, while cotton levies funded research and extension services. In addition, the creation of producer groups that were responsible for handling input supply and cotton marketing at the village level, was another key factor in the rapid increase in cotton production during the 1970/80's.

In order to tackle the monopolies of parastatal companies, the World Bank and other donors have, since the 1990s, put a lot of pressure on national governments to reform cotton sectors (Badiane *et al.*, 2002). The state is gradually withdrawing from chain operations (e.g. privatisation of input supply, ginning and cotton marketing) and price setting now includes state bodies (i.e. ministries or Cotton Boards), private companies and cotton farmers. Cotton producers are thus facing new responsibilities and consequently have begun to organise themselves in farmers' organisations (FO) at district, provincial and national levels.

FO responsibilities go beyond mere price negotiations, as FOs also represent their members in national platforms (e.g. Cotton Boards[82]) and local platforms (e.g. district committees) where policy discussions take place and support services are planned. Cotton sector reforms in French speaking West African countries share a drive to maintain the merits of an integrated approach in the liberalised sectors. Cotton farmers' organisations are now linked to other stakeholders throughout the cotton supply chain, as well as service providing networks - a position which puts them at the heart of the sector. However, the international

---

[82] In French: "Inter-professions".

crisis under which the cotton sector has been suffering since the end of the 1990's, makes the reforms a delicate operation.

Slowly, cotton farmers' organisations have established themselves on the cotton scene and have had to address various issues. In this chapter, we illustrate how cotton farmers' organisations deal with three different issues and the challenges they face: (i) *trade policy and negotiation*; (ii) *domestic cotton price setting*; and (iii) *innovation for cotton growing*. These examples provide direction for the effective empowerment of emergent producer organisations.

2. Farmers' organisations in West African cotton sectors: from state supervision to member-led organisations

In West Africa cotton is mainly grown in French speaking countries (see map below): it provides between 20-50% of export earnings for countries like Benin, Burkina Faso, Chad and Mali. About two million farming households produce cotton, which means that more than 20 million people rely on it directly, irrespective of related secondary activities. Due to the increased liberalisation of cotton sectors, farm gate prices for both inputs and cotton are more closely linked to world market prices, without state subsidy buffers. A rise or fall in cotton prices makes a big difference to the incomes and welfare of millions of people in West Africa (Toulmin, 2006).

The success of West Africa cotton is primarily due to smallholder farmers who worked to create village associations in the 1970/80's (e.g.

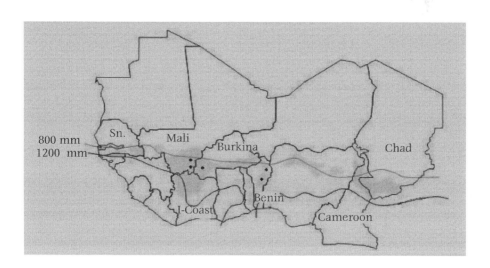

Producer organisations and market chains

the case of Mali, see Tefft 2004). This remains the basis and strength of the cotton producers' movement in West Africa. Initially created to handle logistics, village associations rapidly became key players in community development. A guaranteed price for cotton and rebates[83] provided village associations with important financial resources that were used to fund community infrastructure. Cotton was considered a motor for rural development. In addition, association members benefited from literacy training and management support, which meant that the associations were often employed to manage other community activities not directly related to cotton growing and marketing.

During the democratisation processes in the 1990's, in line with sector reforms, cotton producers set up local and sub-national farmers' unions and finally national organisations. These organisations followed diverse trajectories which depend on specific contexts and charismatic leaders, with the end result being a broad array of organisational structures. Without seeking to provide a complete overview, we briefly present the situation for the main West African cotton producing countries:

- In Benin, Burkina Faso as well as Cameroon, national cotton producers' organisations such as the '*Union Nationale des Producteurs de Coton de Burkina*' (UNPCB), the '*Fédération des Unions de Producteurs*' (FUPRO-Benin, with its recent branch '*Association Nationale des Producteurs de Coton*' ANPC) and the '*Organisation des Producteurs de Coton du Cameroun*' (OPCC) take part in the chain as shareholders in cotton companies (e.g. the former parastatals in Burkina Faso and Cameroon) and/or stakeholders in Cotton Boards (e.g. Benin, Burkina).[84]
- In Mali, where sector reform started later than in the other countries, cotton producer unions such as the '*Syndicat des Producteus du Coton et Vivriers du Mali*' (SYCOV) arose from village associations that sought representation in cotton price negotiations. Circa 2000, three other syndicates came onto the scene. They are now involved in sector reforms. The '*Association des Organisations Paysannes Professionnelles*' (AOPP) is another strong cotton farmer representative. However, neither the *Syndicats* nor AOPP are operators within the cotton supply

---

[83] A farm gate price is defined before marketing cotton. However, when actual selling on the world market takes place, prices may end up being higher. This allows payment of a separate, supplementary amount - the rebate ("*ristourne*") - to cotton growers.
[84] Benin has history of schisms within farmers' movements; in addition to FUPRO, other dissident organisations claim to represent village associations. Recently the government has created the "*Conseil National des Producteurs de Coton*" which represents all the cotton producers' organisations. In 2005/06 FUPRO launched the "*Association Nationale des Producteurs de Coton*" as a special branch organisation for cotton producers.

chain, nor do they provide economic services (e.g. input supply) to village associations[85]. This determines the strength of their position vis à vis other stakeholders. Since 2004 village associations have transformed into cooperatives and district and provincial unions have been created. In 2007, following the example of Benin and Burkina, the newly created '*Union Nationale des Sociétés des Coopératives de Producteurs de Coton*' (UN-SCPC) takes part as direct stakeholder in the sector and will take over some *Syndicat* functions (e.g. price setting).

The landscape of cotton farmers' organisations in West Africa is quite diverse. Furthermore, these organisations are growing accustomed to positioning themselves on different fronts and at different levels and dealing with a wide range of stakeholders. They share some striking features:

- National cotton producers' organisations are relatively young and still adapting to new roles and functions. They are the apex of multi-tier organisations and have to cover a wide spectrum of issues. A clear definition of core functions at the different levels is required to enhance organisational performance. This also includes addressing the issue of the particular role of cotton and cotton farmers' organisations in rural development.
- All cotton farmers' organisations have their roots within village associations that were created and controlled for a long time by state services. Most national organisations have emerged - and some, like the OPCC of Cameroon still are - under the continuing supervision ('*tutelle*') of parastatals or state services. This makes political cooptation a real-life risk when considering the issues at stake in reform processes. A transformation into genuine democratic, membership-based and member-led umbrella organizations remains a key challenge.[86]
- Most support programmes in countries like Mali, Burkina Faso, Benin or Cameroon, continue to bypass state departments and/or are sub-contracted to agencies (consultancy firms, NGOs), which often have

---

[85] Producers' organisations (Syndicates) were responsible for non-cotton input supply for four years, but CMDT still gave the financial guarantee and organised distribution.
[86] Some farmers' organisations are in significant debt, e.g. in Burkina Faso UNPCB experiences financial trouble as a shareholder; www.jeuneafrique.com/pays/burkina/article_depeche.asp?art_cle = AFP31357lafilsyapsa0.

Producer organisations and market chains

their own perceptions and agendas. Support programmes are rarely negotiated directly with national and local farmers' organisations.

3. Trade policies: from un-fair trade to fair trade?

Unquestionably domestic support and subsidies from the United States, the EU and China keep the world price of cotton-lint down. The story is well known: in the United States 25,000 cotton farmers enjoy 3 billion USD in subsidies, and the EU and China support their production each with about 1 billion USD. Two million farmers in West Africa receive no subsidies whatsoever, but suffer the consequences of lower world market prices, which without those subsidies would be 10-15% higher, and thus would provide 20-40 million USD more each year in each country. Farmers would receive 30% of this increase (Gillson *et al.*, 2004; Goreux, 2004). The cost-effectiveness of West African cotton producers is even further influenced by the 'strong' Franc CFA (the common currency for French speaking countries in West Africa that is directly linked to the Euro) and higher input prices (mostly petrol based). In response to the demands of cotton producers, the governments of Brazil as well as Chad, Benin, Mali and Burkina Faso have challenged subsidies within the World Trade Organisation (WTO). National governments are therefore key allies for cotton producers' organisations in their battle against disturbing subsidies as they share a common interest: revenues for national economies and farm households. West African governments and farmers' organisations have a strong case - the more so given that Northern countries are violating their own rules. However, to date the case remains unsolved at the expense of cotton farmers in West Africa.

Yet, the national producers' organisations decided to go further by joining forces at the continental level. Organisations such as the '*Réseau des Organisations Paysannes et de Producteurs de l'Afrique de l'Ouest*' (ROPPA; created in 2000) and the more specialized '*Association des Producteurs de Coton Africains*' (AProCA; created in 2004), are emerging as two important international platforms for national farmers' organisations. Generally, the global abolishment of cotton subsidies is considered to have positive effects on West African cotton sectors, but the WTO is a dangerous arbitrator: cotton is connected to other agricultural products, like food crops, and West African countries may have to open their markets to US, EU and Asian products as a 'fall-out' of WTO lobbying. However, the West Africans do not have much of a choice. The WTO is the only legitimate platform, and is still preferable to bilateral Economic

Partnership Agreements (EPA), which most countries risk signing after the recent WTO failure. Furthermore, cotton is one of the few cases where Africans can beat Northern countries with their own weapons: West African cotton sectors were, and probably are, competitive and viable. It is self-evident that the USA and the EU are protecting their own dwindling cotton sectors to the detriment of millions of people in West Africa.

However, in the discourse on farmers' organisations there is no contradiction between cotton lobbying for abolishing subsidies and demanding a fair price for export cash crops, versus the lobby for protection of food products on the domestic market, by defending import barriers (for example, the principle of 'food sovereignty', which almost every industrialised country has applied in the past). The lack of protection of the domestic market - a well-known 'urban bias' in politics - is contested by ROPPA, which took a clear stand on 'food sovereignty' from a livelihood perspective thereby voicing the concerns of their membership; West African family farms.[87]

AProCA, as a particular platform for cotton farmers' organisations, took the trade issue further. It started exploring possibilities for 'alternative cotton markets' such as organic cotton, Fair Trade cotton and West African sustainable cotton. A first step taken by AProCA was to inform member organisations of these possibilities and to share member experiences with these niche markets. The potential for these cotton outlets is very promising. However, they are, and remain for the moment, 'niche markets' that cannot meet the needs of the majority of cotton farmers.

4. Domestic price setting: from monopoly to transparency?

Besides lobbying for the abolition of distorting subsidies (on the agenda since AProCA's creation), price-setting is another major issue for AProCA and national unions. Cotton farmers in countries like Mali and Burkina Faso are accustomed to and attached to the concept of minimum prices, which are announced in beginning of the growing season allowing production capacity and necessary investments to be planned. Price setting is another battlefield between the state, cotton parastatals, farmers' unions, banks and donors. In the 1980/90's cotton prices for farmers were too low for too long, with farmers gleaning less than 50% of

---

[87] See: ROPPA: *Appel de Niamey pour la Souveraineté Alimentaire de l'Afrique de l'Ouest.* Available at www.roppa.info/IMG/pdf/Appel_de_Niamey_definitif.pdf

the world price for cotton-lint, which was also the case in 1995-98 period. Since 2000 the situation has improved; farmers receive 60% or more of the world price (Badiane *et al.*, 2002) with the remaining amount going to the ginning industry, taxes and stability funds, exploitation costs, overhead - and sometimes mismanagement - of cotton parastatals. On the other hand, since 2000 parastatal companies withdrew from several public services, which have not been taken over by states. Farmers are no longer protected from yearly fluctuations, for which stability funds used to be in place. Moreover, most countries have difficulties in fuelling the stability funds when prices are low prices (Baffes, 2004).

In increasingly liberalised cotton sectors, in order to get a more realistic reflection of the market, prices paid to cotton growers are linked to the world price. However, the world price itself is currently an outcome of the imperfect functioning of the international market due to distorting subsidies. Furthermore, cotton companies and government departments still set the rules for the prices paid to farmers, and when necessary, combine it with political pressure on farmer leaders to accept their propositions. In addition, cotton parastatals tend to be reluctant to provide information to farmers' organisations as to levies and the functioning costs of the cotton supply chain. Lack of information and insufficient 'market intelligence' make it difficult for farmers' organisations to join the power play. AProCA supports its members in price negotiations because price-setting mechanisms and negotiations only work if all parties have the same level of information. AProCA therefore has paid particular attention to, and improved, information sharing and joint analyses on cost prices, operational costs within the cotton supply chains, etc. among national member organisations.

The ANPC/FUPRO of Benin, for example, benefited from preparatory consultations with counterpart organisations in Cameroon, Mali and Burkina Faso and was able to ensure a 2006 price that was at the same level as neighbouring countries; around 170 FCFA/kg as opposed to less than 150 F/kg.[88] Sharing experiences between national and local farmer leaders allows for mutual understanding of the tricky business of price setting and stabilisation as well as a better understanding of the stakes involved. In Mali, the national AOPP moderated debates and published results in national languages (for all 6000 member cooperatives) in order to facilitate exchange between member farmers' organisations; for example farmers write their own articles in the journal '*Jekaa Bara*' (over

---

[88] Concerning about 200 000 tonnes of cotton for 300 000 households. Personal communication with FUPRO.

6000 copies each trimester). AOPP also has a technical committee that monitors cotton sector reforms and proposed, for example, a stability fund supplied by financial resources from farmers' organisations in order to maintain a minimum cotton price for all producers.[89] They hope that this farmer-led initiative makes the national government and donors keep their promise of fund provision.[90]

The 2007 price negotiations in Burkina Faso and Mali confirm the need for information sharing. They also proved to be revealing for the position of farmers' organisations with different historical backgrounds. Because of the low world market price for cotton and a low exchange rate on the Dollar, the farmer price in Burkina Faso was set around 145 FCFA/kg, which is an historically low price (the price was at 160 and 175 in 2006 and 2005 respectively). In Mali the farmers will receive 160 FCFA/kg, as in 2005. The world price is not the only reference point; price setting is also a national political decision, which farmers' representatives have been involved with for 15 years.

The UNPCB was the only farmer stakeholder in Burkina Faso Cotton Board negotiations. As a shareholder in cotton companies, having economic responsibilities within the chain, it found itself in the delicate position of defending farmers' interests as well as the financial interests of companies. Almost no preparatory debates took place between farmers and their representatives; the UNPCB confronted their members with a 'fait accompli'. In Mali farmers' leaders of the UN-SCPC, backed by members of the syndicates, resisted government and company pressure. They stuck to the minimum price, in accordance with the 2005 agreement. Exchanges between local unions in Mali and Burkina Faso made farmers in Burkina Faso realise what was happening; they started to pressure their national Union to change farmer prices for cotton and inputs.[91]

5. Innovation for cotton growing: from intensification to sustainability?

Increasing cotton production over the past three decades in West Africa is undeniably a success story of improved farming systems. This increase

[89] Undertaken with technical and financial support from the Netherlands Development Organisation SNV.
[90] The price went from 160 FCFA/kg to 165 or more for 500 000 tons of cotton produced. Price stabilisation is also crucial for keeping production at realistic levels and sustains input supply for all farmer households in South-Mali.
[91] They succeeded in lowering input prices at a value of 6 billion FCFA (Pers. Communication with farmers, April 2007).

is mainly the result of area extension for cotton and cereals, but was at the expense of forests, pastures and fallow land, thus endangering the traditional method for soil regeneration. Yet, increasing yields of both cotton and cereals are also the effect of a more intensive use of inputs, such as fertilizers and pesticides, labour and agricultural equipment. This was made possible by appropriate credit facilities and an adequately coordinated and dense network of research and extension services. The investments needed were ensured by the cotton revenues of farm households and cotton levies for funding services (Zoundi, 2005).

Under current reforms, service provision is changing radically. The overall trend is that private goods such as agricultural inputs and related technological knowledge are increasingly provided by the private sector[92] while more complex issues such as soil fertility management are addressed by public sector services, which suffer from an almost persistent lack of human and financial resources. Sustainability of production systems is now high on the (farmers') agenda and numerous initiatives to address the key issue of soil fertility management are underway. We think that cotton-based farming systems are on the brink of an agricultural revolution, which began with intensification in the 1990's: an increase in the number of cattle (and in manure production), in equipment and in the use of chemicals fertilisers.

However, scaling up this intensification towards a more sustainable agriculture requires several critical issues to be addressed including: low cotton prices; insecure land rights; inappropriate credit facilities (outside the cotton supply chain) preventing farmers from investing in land management; and, limited access to research and advisory services, which keeps farmers from having up-to-date information. These issues are on the agenda for all national cotton farmers' organisations.

Still, farmers' organisations must also address members' service needs. For example several local cotton producers' unions in Burkina Faso and Mali are experimenting with a farmer advisory service that includes land use planning, resource allocation and bookkeeping, the so-called the '*conseil agricole paysan de gestion d'exploitation*'. This farm management approach was developed in the 1990's by several research institutes. However, it faced problems in terms of application by cotton farmers due to the significant human and financial resources required for which state extension services and donor-funded projects were often solicited. In 2001 district cotton farmers' cooperatives of the Tansila

---

[92] The introduction of genetically manipulated varieties is a subject of debate (e.g. tests in Burkina Faso with Bt-cotton).

department, in the Banwa Province of Burkina Faso, decided to access services for their members. Although located in the heart of the cotton basin, Tansila was a remote area that was poorly serviced by the state and donor-funded projects. The departmental union, which federates 70 cooperatives, decided to adapt the farm management approach for its own purposes. With initial support from development partners[93], the union simplified and translated manuals and information sheets into the local language (Dioula). The availability of literate and trained young people made it possible to set up a network of farmer-trainers and farmer-supervisors, entirely funded by the cotton revenues generated by the union.

An evaluation in 2006 showed remarkable results considering that funding for state extension services was limited and that union cotton revenues had gone down. More than one third of the 70 village cooperatives now use the farm management approach. In these cooperatives overall cotton yields rose and farming systems have more balanced rotation plans (wherein food security was ensured) than in the other cooperatives; there is also evidence of better family revenue management, meaning that the household head paid more attention to the needs of women and young household members than in the past (Traoré *et al.*, 2006).

Box. Testimonies on use of management device in Tansila

'For me the 'conseil agricole paysan' has changed my life. I was always convinced that I should get a lot of money by just having large surfaces of cotton. Last year I have reduced cotton from 24 to 15 ha and I got more 23 in stead of 15 Tons by adding good manure, thanks to the help of the young advisors. In fact, we only cultivated for the Bank and the cotton company, since revenues were very meagre compared to the costs. This work has changed in four years what 27 years of state extension could not achieve'... 'The objective is no longer to be a big cotton producer, but to have good rotations of crops, based on three principles: food sufficiency (not being obligated to sell cereal stocks in rainy season), reasonable debts (for inputs) and possible revenues. Food is most important, cotton affords us to have inputs. After that we look for speculations that give us complementary revenues, like sesame'... 'The 'conseil agricole paysan' has allowed a lot of families to get out of the nightmare of lack of food and high debts'

---

[93] A team from the Netherlands Development Organisation SNV and local agricultural extension services.

In 2007 this farm management approach will be used by other cooperatives in Mali and Burkina Faso. However, these innovations, which are only supported by local FO, are marginal at the national scale. In the past several donor-related agencies have developed approaches in national programs, but no program sufficiently took into account the importance of institutional sustainability at local levels. The question of scaling-up is topical; 'connecting' different initiatives and programs is a first major step.

6. Concluding remarks: the empowerment of cotton farmers' organisations

Today, cotton farmers' organisations are primary stakeholders in the cotton sector, which remains a key sector for national economies and smallholder farm households in West Africa. These farmers' organisations are operating at different levels and dealing with a wide spectrum of issues, as illustrated above. The three examples we offer, illustrate how farmers' organisations, at their own initiative, have gradually come to grips with some rather complex issues in the cotton sector. These examples also provide us with some direction vis à vis supporting cotton farmers' organisations in strengthening their position as legitimate representatives of smallholder cotton farmers, and contributing to successful reforms in the cotton sector to benefit cotton farmers.

First of all, under pressure from Bretton Woods institutions, economic functions have been transferred from parastatal companies to the private sector and state services. Cotton reforms made the sectors more transparent and offered new opportunities for farmers' organisations. However, neither the State, nor the FO received the necessary support to face the huge task of reorganising a sector, which is still most important in countries like Benin, Burkina Faso and Mali. Ever-changing decisions and donor support bring more instability in already vulnerable sectors. Therefore coherent, long term, nationally owned, participative and well-financed 'cotton' policies remain a prerequisite for successful reforms.

Secondly, most current support for cotton farmers' organisations happens indirectly through other actors (line ministries, donor-related agencies, etc.) as part of cotton sector reform programmes. In our view this is an obstacle to the 'social construction' of member-led, autonomous farmers' organisations that then have to confront other actors. Support therefore needs to be designed and implemented in close collaboration with farmers' organisations, and more importantly, should involve long-term commitment from all actors involved.

'Empowerment' should be the leading principle for support to cotton farmers' organisations. As member-based organisations they must develop the capacity to make strategic choices and undertake actions in a sector where this was previously denied. Therefore support goes beyond mere management support to include the governance of both member-based organisations and the cotton sector since new rules and regulations have to be designed. In our view, support to member-based, autonomous organisations means becoming partners and critically accompanying these organisations by asking relevant questions rather than providing 'blue prints'.

For example, the Netherlands Development Organisation (SNV) supports cotton farmers' organisations in Mali, Burkina Faso and Benin, where it has developed partnerships at all levels of intervention including: with AProCA and national platforms to better prepare for negotiations on trade policies or cotton reforms; and with local farmers' organisations and cooperatives in order to improve service delivery to members. This leads to the third direction: appropriate support requires various competencies and adequate coordination. Several Dutch development organisations (e.g. Agriterra, SNV, ICCO, Solidaridad and KIT) have taken up this challenge and now work directly in partnership with farmers' organisations. They recognize the need for long-term commitment and grasp the complexity of the situation. They have joined forces in the Agri-ProFocus platform, in order to stimulate knowledge exchange and enhance the effectiveness of support.

References

Badiane, O., D. Ghura, L. Goreux and P. Masson, 2002. Evolution des Filières Cotonnières en Afrique de l'Ouest et du Centre. Banque Mondiale/ Fonds Monétaire International, Washington DC Available at http:// www-wds.worldbank.org/external/default/WDSContentServer/IW3P/ IB/2002/10/12/000094946_02080604014034/additional/105505322_ 20041117162010.pdf

Baffes, J., 2004. Cotton, Market Setting, Trade Policies and Issues. WB Policy Research Working Paper 3218. Development Prospects Group, World Bank, Washington DC, Available at http://www-wds.worldbank.org/servlet/ WDSContentServer/WDSP/IB/2004/06/03/000009486_20040603091724/ Rendered/PDF/wps3218cotton.pdf

Gillson, I, C. Poulton, K. Balcombe and S. Page, 2004. Understanding the Impact of Cotton Subsidies on Developing Countries. Working Paper, ODI/ DFID/ Imperial College, London. Available at http://www.odi.org.uk/iedg/ Projects/cotton_report.pdf

Goreux, L., 2004. Prejudice Caused by Industrialised Countries Subsidies to Cotton Sectors in Western and Central Africa. Mimeo, second edition, France.

Tefft, J, 2004. Mali's White Revolution: Smallholder Cotton from 1960 to 2003. Paper presented at the NEPAD/IGAD regional conference 'Agricultural Successes in the Greater Horn of Africa', Nairobi, November 22-25, 2004. Available at http://www.anancy.net/uploads/file_en/p10-cotton.pdf

Toulmin, C, 2006. Rachel Carson Memorial Lecture, PAN, UK.

Traoré, N., J. Sombie and M. Badiel, 2006. Les Organisations Paysannes en Zone Cotonnière, Peuvent-elles Fournir un Service de Conseil Efficace aux Exploitations Familiales? Exemple du Conseil Agricole Paysan Conduit par Deux GPC et l'UDPC Tansila au profit de leurs affiliés. SNV, Bobo-Dioulasso, Burkina Faso.

Zoundi, J.S., 2004. La Transformation de l'Agriculture Ouest Africaine: Vers des Nouveaux Partenariats pour l'Innovation Agricole. Etudes de Cas sur l'Innovation Agricole et le Sous-Secteur Cotonnier au Mali. SWAC/OECD, Paris.

Section C

Changes in the institutional environment for producer organisations

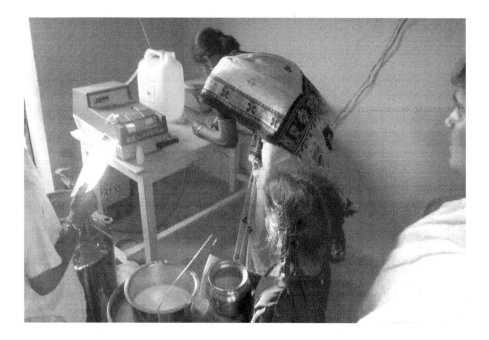

The farmers' organisation route to economic development

Kees Blokland and Christian Gouët

This chapter highlights Agriterra's perspective on the contribution of farmers' organisations to economic development. Agriterra is a Dutch organisation supporting international cooperation amongst farmers' organisations. The authors argue that farmers' organisations make a positive contribution to economic growth, democratic relations, income redistribution, and poverty reduction.

1. Introduction

This chapter describes the analytical framework on this subject that has shaped Agriterra's development cooperation practice. The perspective is summarized in Agriterra's axiom: farmers' organisations make a positive contribution to economic growth, democratic relations, income redistribution, and the reduction of poverty. This is the impact that we (at Agriterra) work towards in our development cooperation. As such,

Agriterra's direct goal in development cooperation is to strengthen farmers' organisations.

2. Viability of a Via Campesina

The idea catalysing the formation of Agriterra was the that development could be structured so that the dominant influence came from the peasantry (Blokland, 1992)[94]. It was argued that prior to that point the social sciences, when exploring a via campesina[95] to development, had produced convincing social criticism, as well as a body of evidence demonstrating the micro-economic advantages of small farmer productivity. However, this was not enough to persuade policy makers to consider the contributions of farmers in economic development. Neither scientists nor policy makers could imagine how development - improvements in productivity and living conditions - could happen while farmers remained a dominant social grouping. The fact is that development involves an ongoing reduction in the number of farmers. Initially farmer population reduction is relative - their representation as part of the economically active population. But ultimately the decreased number of farmers is absolute: farmers constitute a very small minority in developed countries.

Blokland's study of the Nicaraguan National Farmers' and Ranchers' Union (UNAG) in the 1980s elaborates on the relationship between the organised peasantry and development (Blokland, 1992)[96]. The study develops an argument building on four concepts that constitute the socialist theory of economic development: participation, accumulation (economic growth), socialization of the means of production, and planning. Examining the positions of different actors and development practice in Nicaragua at that time (state versus farmers' union), shows that the involving influence of the organised peasantry in development leads to new interpretations of these socialist concepts. Together these new interpretations provide an arrangement that differs substantially from socialist orthodoxy. As these interpretations are based on everyday

[94] Please note that the term "peasantry" is used as a concept to encompass all rural producers who derive an income from tilling the soil or herding, including small, medium and large scale farmers.

[95] Via Campesina (in English: *peasantry path*) is the Latin American contribution to theories on development paths envisaged by Vladimir I. Lenin (Farmer Road and Junker Road).

[96] After Zamosc (1987), Blokland's 1992 study was the second thorough examination of a farmers' union in Latin America. Biekart (1999) published a third study on ASOCODE as a chapter in his PhD thesis.

practice of the UNAG during the Sandinista transition to socialism they underscore the viability of this route to development for many agrarian economies in the developing world. Hence, the conclusions of the study go beyond the transition to socialism and present the case for a Via Campesina[97]; a possibility that seemed even more likely to occur after the Sandinista era, when farmers welcomed the market liberation policies of the government that took office in 1990 (Blokland, 1996).

This peasant development path is built on the full participation of the peasantry in decision making and development planning, together with government and other social sectors, at all levels (local, national, international) depending on the scope of the organisation. It allows development to take place not only with the material contribution of small farmers to (state-led) industrialisation, but with farmers as key players in the structural transformation of the economy: the transition of a predominantly agrarian economy into a modern one with manufacturing and service industries.

This path does away with prescribed socialist models of farmer organisation for production (collective farms), which in the past were promoted by development cooperation. At the same time, the study shows that it is advantageous for farmers to organise, so as to enable the introduction of new technology in agriculture. This is an important confirmation of socialist expectations. The association of farmers, including through collective farms fosters debate and enables field experiences. However, it is not so much collective farming itself, but collective reflection that helps with the introduction of technology. The so-called 'peasant collectivization' implies recognition of the force of collective decision making and negotiation with agricultural research, not only at the local level, but also at higher levels in the federation of farmers' associations (see also IFAP, 1995). There is evidence that organised farmers are at an advantage in terms of getting to know more technological options, because they can bring together options from different regions within a country and from other parts of the world. Yet, the adoption of new technology is an individual decision and its implementation is the responsibility of the farming family.

The development path that arises under farmers' influence is more labour intensive as industrialization and the development of services take place in decentralized farmer-led enterprises close to primary production. It does not, however, alter the fact that development leads

---

[97] Blokland was at that time working on the establishment of an international platform of rural people's organisations named 'Via Campesina'.

to a reduction in the number of farmers, because that is inherent to increasing productivity and living conditions. Farmers and their families start working in the newly established enterprises that emerge as a consequence of their own initiative. This structural transformation of the economy is expected and once it happens it acts to safeguard employment and food security. Thus, the Via Campesina path has far fewer disruptive social effects than development paths with heavy state intervention or alternatively under completely free market conditions. At the same time it allows farmers to continue to play a dominant role in development, contributing to planning, absorbing investments and building new ventures. Governments are not convinced of this route, but likely support it as an outcome of social negotiations between government and social sectors. The government's role is to strengthen weak players in the market and create a more level playing field.

Development cooperation also has a contribution to make to this development path: to strengthen farmers' organisations in taking positions vis-à-vis government and other social players and for assisting them in developing a 'peasant-based mode of appropriation'. This latter concept refers to all farmer organisations at local, provincial and national levels, in order to get control over resources for development that match the farmers' ideals. Following the example of Nicaragua, these ideals emerged from massive participatory development planning processes that were conducted at the time by UNAG with its members.

The arguments introduced above on farmers' organisations and development can be summarized by the following elements:
- Development with a dominant influence from farmers is conceivable when the farming sector becomes a full participant in development planning, net receiver of new investments and entrepreneur in new ventures for processing and servicing agriculture.
- The basis of full participation is farmer unity from local associations up to the international level, in organising for negotiation with other sectors and government, and in developing a specific peasant mode of appropriation.
- Collectivization of primary production has, in some cases, advantages in the early stages of development as long as conflict with the

landowning class exists, but will become counterproductive once farmers opt for development (stage theory of collectivization[98]).
- In the early stages of cooperative development social-political and even military objectives of the cooperative prevail over economic ones.
- Advantages attributed to the collectivization of agriculture for development (the smooth introduction of new technologies, improvements in income distribution and faster economic growth) can be attained without sacrificing private ownership or individual tillage. 'Peasant collectivization' refers to collective preparation for technological decisions (Blokland, 1992).
- Advantages of market-oriented reforms that break away from state interventionist policies are appreciated by farmers with secure land rights and investment plans, i.e. by entrepreneurs (even very small entrepreneurs). Market-oriented reforms lead to a second stage in cooperative development with the service cooperative being the predominant model (Blokland, 1992, 1995, 1999a.).
- Service cooperatives (second stage cooperative development) emerge once economic conditions are favourable and economic development is taking off, when income distribution is improving and when modern technology is being introduced.
- In this second stage, it is not wise to establish cooperatives with only resource-poor members. Social heterogeneity is a basis for economic success for cooperatives (van Diepenbeek, 1990; ICA, 1977).

3. The emergence of farmer-led development cooperation

From these contributions to theory and the subsequent lobby for agrarian development cooperation, a practice of support to farmers' organisations in developing countries emerged. In the mid-1990s, the Dutch NGO PFS[99] was reorganised in order to build the farmer-led development cooperation agency, Agriterra. Agriterra is an 'agri-agency'. Agri-agency is a term coined by Agriterra that refers uniquely to development agencies that are steered by rural producers' organisations, cooperatives and other rural membership based organisations in the home countries

[98] This is not to say that all countries should pass through a collectivization phase, but cases where collectivization of agriculture, in the classical sense of the concept, is functional are those where a skewed distribution of assets need redistribution for the creation of a more level playing field.
[99] Paolo Freire Foundation

(in developed countries) that, and because they exclusively work with similar organisations in the developing countries (See Box 1). Agriterra was established because the Nicaraguan experience determined that not only financial support but also practical solidarity particularly from Norwegian and Swedish farmers helped UNAG to take a stance that was more independent from government at the time. In the 1990s Agriterra assisted this union in its effort to establish a rural bank, following the

Box 1. Agri-agencies

Agri-agencies are 'NGOs' for Development Cooperation structurally linked to farmers' and rural members' organisations. All are funded or steered by organisations of farmers, rural women, young agrarians, co-operatives and agri-businesses members and representatives. They only co-operate with other farmers' and other rural membership based organisations in developing countries; rural organisations that are accountable to their members.

AgriCord, The Alliance of Agri-agencies. Its mission is a reflection of the missions of its members, and can be summarised as: 'To strengthen farmers' and rural membership based organisations in developing countries and countries in transition'. In living up to this mission, AgriCord and its members focus on three main strategic issues:
• Provision of more and better services to members;
• Enhanced role in civil society to defend the interests of farmers and rural members;
• Development of economic activities to support members.

AgriCord's strategy can be summarised by the following points:
• Agri-agency specific cooperation approach: work only with farmers' and rural membership organisations and their co-operative/economic initiatives.
• Cooperation and partnership based on reciprocity between farmers' and rural members' organisations internationally.
• Coordination of programmes with the aim to further specialise agri-agencies by dividing tasks and work.
• Partnership with IFAP-Development Cooperation Committee, and facilitation of direct contacts between farmers and rural members' organisations within the platform.

More information: www.agricord.org; http://nl.wikipedia.org/wiki/Agri-agencies.

outcome of a participatory strategic planning exercise that had involved tens of thousands of cooperative members in Nicaragua. Agriterra, used this approach again for organising participatory planning processes in South American farmers' organisations in a programme called PIPGA, which was supported by the Dutch Ministry of Development Cooperation (see Biekart, 2000), and more recently formulated it as a worldwide work area called Participatory Policy Formulation (Agriterra, 2006).

Likewise, Agriterra promoted the involvement of other farmers' organisations from Central and South America, the Caribbean and Africa in farmer-to-farmer exchanges on agricultural technology. Agriterra's efforts directly build upon the potential benefits attributed to peasant collectivization that are part of these exchanges.

In a similar vein vis à vis farmer cooperation, Agriterra's Annual Report 2000 states the following: 'In its vision on the future, the management of Agriterra sketches for the year 2010 a world-wide organisation for agrarian development founded upon national agri-agencies. On its path towards this aim, Agriterra wishes to consolidate its specific way of working in close collaboration with other agri-agencies [see Box I] the members of the Development Cooperation Committee of IFAP will steer the program, while the agencies will be responsible for the implementation' (Agriterra, 2000: 6). Hence, Agriterra promotes the organisation of farmers at all levels, especially regional (supranational) and international, as a lot of work needs to be done at those levels. In fact, since its launch in 1998, Agriterra promoted alliance with its sister agri-agencies, which culminated in the establishment of AgriCord, as it was officially launched in 2003. Later, this alliance joined efforts with the International Federation of Agricultural Producers, IFAP (Gouët, 2004). As a consequence of agri-agency involvement IFAP membership increased from 81 in 1992 (27 from developing countries) to over 110 national farmers' organisations in 2006, more than doubling developing country members. Currently, the agri-agencies strengthen capacities for over one hundred national farmers' organisations and support them in establishing new economic ventures in production, processing, trade and services to agriculture. This inclusion of smallholder organisations in IFAP, the creation of AgriCord and its strategic partnership with IFAP-DCC responded to the basic principles of Agriterra (Gouët, 2004).

For development cooperation through agri-agencies is clarity as to what strengthening of farmers organisations means, is crucial. From a theoretical point of view Agriterra's main contribution to the theory of institutional development and organisational strengthening is that capacity building in a farmers' organisation must be in line with and targeting the member services it plans to develop; services that must match the mission of the organisation (Blokland, 1999b).

A second contribution is that Agriterra developed a method (the 'profiling method') to measure how much the capacities of rural producers' organisations are strengthened. The profiling method refers to the collection of data and opinions vis à vis a farmers' organisation and the development of leading indicators. From recurrent profiles of the same organisation, undertaken at three years intervals, one can derive the aspects strengthened and even measure the progress. From the many profiles of rural producers' organisations done so far, a tendency towards stronger organisations can be derived: organisations with more members and member participation, professionalism, accountability to members and outsiders, outside representation, strategic potential and a better gender balance.

Once strengthened, the impact of farmers´ organisations on development, democratization or income distribution is taken for granted. Impact cannot be derived from agri-agency practice. In 2006, 45 million Euro in donations to farmers' organisations in developing countries was managed by agri-agencies, but this does not suffice to show development impact. Then again, within and outside of Agriterra the demand for more evidence as to the impact of its work on development is increasing. The newly launched programme `Farmers Fighting Poverty´ includes efforts to come up with examples of plausibility, using the method of Harvesting of Stories[100] and Most Significant Change (Dart and Davis, 2003), amongst others. This is not an optimal solution, but a first step towards evaluating development impact. Notwithstanding these additional efforts, so far we have only looked at the issue from a theoretical standpoint and will now summarize the insights to date.

4. Farmers' participation and development

Inquiring as to the relationship between farmer organisation and development, and supported by scientific evidence from research undertaken at the Fraser Institute and by Pamela Paxton (Paxton,

---

[100] *Harvesting of Stories* is a method coined by Paul Engel (ECDPM) and further developed by Christian Gouët (WU). It was introduced for the first time during the evaluation of Agriterra's program 2001-2003.

2002)[101], the following contributions to the theory are recorded in Agriterra publications:

- The causal relationship from the association of farmers to economic development runs via the improvement of democratic attitudes and relationships fostered by open associations, i.e. associations whose members maintain positions in several other civil society organisations.
- Not every farmers' organisation will foster economic development but only those which are open to the outside world and opt for 'concertation' (Paxton, 2002; Blokland, 1995, 2002, 2003; Schuurman, 2006).
- The proven relationship between economic freedom and development implies that countries with a low ranking on the Human Development Index and at the same time enjoy high economic freedom during long periods will display a favourable environment for development cooperation activities (www.freetheworld.com, see also Blokland, 2000).
- However, from the lack of, or even inverse relationship between economic freedom and development in the lower quintile of countries (ordered according to GNP), a stage theory of economic development can be deduced. This suggests that state interference in economic activities at an early stage could be beneficial for development. The benefits can be drawn from examination of the first stage of collectivization during the implementation of land reform. Forced land redistribution is a limitation to economic freedom, while it creates at the same time many more property owning farmers and sets the stage for economic development (Blokland, 2002).
- The positive relationship between farmers' organisations and development can be viewed taking into account the drive for political democracy and economic freedom (Blokland, 1995).
- Beyond the biased agricultural policies that prevailed throughout most of the 20th century (Schiff and Valdéz, 1992), new inequalities emerged in the agricultural sector between a small sector of dynamic entrepreneurs and a large sector of marginalized campesinos (Spoor, 2002).
- The new economic paradigm, beyond state interventionism and liberalism, is economic democratization (Blokland, 2003).

---

[101] See also www.fraserinstitute.org

Producer organisations and market chains

5. New contributions

Theory and empirical evidence from a range of different perspectives suggest and confirm that farmers' organisations and other rural membership based organisations can indeed be relevant actors in development and poverty alleviation in rural regions (Uphoff, 2000; Rondot and Collion, 2001; de Morré, 2002; Uphof and Wijayaratna, 2000). Research at Wageningen University focussing more specifically on rural producers' organisations (RPOs) and their federative bodies like IFAP, confirms basically what scholars and practitioners have previously noted (Gouët, 2004). On the one hand, actors recognize changes in development were promoted or facilitated by RPOs, for example: increased productivity enabled members to access technological packages; improved marketing occurred because RPOs provided solutions with economies of scale; and, many other examples of development benefits can be attributed to the collective action of farmers. On the other hand, theory suggests that the main contribution of rural producers' organisations for development is the opening up of possibilities for networking, information exchange and exposing individual members and member associations to other actors. In cyber terminology, RPOs are at times vehicles for 'downloading' support, initiatives and contacts to the local level and at other times vehicles for 'uploading' local farmer needs, visions and initiatives to decision making spheres, investors or foreign development agencies.

In line with the above, but from different perspectives, a number of scholars point out three kinds of relationships that RPOs maintain: (1) relationships with members; (2) relationships with state and civil society; and (3) relationships with market players (Engel, 2002; Stockbridge *et al.*, 2003; Presno, 2004). The latter is particularly true when the RPO is dealing with trade issues (even when it does not participate directly in trade, but rather in policy making). Indeed, in several international seminars and workshops with the participation of RPO leaders and officers, the need to improve networking capacity with companies and strong market parties and, policy makers on trade issues, both nationally and internationally, stood out. These seminars included the following topics: small holder marketing cooperatives (one in Asia and one in Latin America), the role of RPOs on contract farming (mainly with representatives of Africa), women participation in Latin American RPOs (Gouët, 2004; Quiroz *et al.*, 2004; Stessens and Gouët, 2004; IFAP, 2005). In all of these seminars, participants noted a crucial role for their RPO was that of networking. Likewise, Gouët and van Paassen. (unpublished) interpret this need and added a fourth kind of relationship to the list above: relationships with peers from all over the world. This additional kind of relationship highlights the importance of working through a peer-to-peer approach and learning from and with other colleagues (see also chapter 4).

This finding derived from practice is coherent with current research could add in terms of explaining potential roles for RPOs in the promotion of development[102]. Gouët's research will bring a new perspective to the issue. From different theoretical positions like innovation theory and social dilemma theory, it can be argued that creating transparency in information amongst diverse actors and enlarging their networks are crucial steps towards generating trust, fostering innovation processes and overcoming social dilemmas. As a result, facilitation of economic and human development can be expected. A peer-to-peer approach for rural producers' organisations support is therefore a crucial piece of the development puzzle. To further analyse and develop methodologies for implementation and evaluation is to take seriously both scholarly findings and the proposals of farmer leaders and rural producers' organisation officials.

But the puzzle remains. A lot of research still needs to be done in order to determine the types of organisations and under what circumstances, and through which actions, can better economic development actually be supported. If we accept that RPOs have a role in development, what then are the basic features within an RPO that need to be supported and even encouraged? These are questions still lack unique and convincing answers. Nevertheless, there are lessons to be learned both from experience and from science. From different perspectives and in diverse specific situations, RPO contributions to development appear to be related to the fact that associations makes it possible for farmers to exchange ideas, experiences, technical knowledge[103], marketing information, contacts and so on (Gouët and Leeuwis, 2004a,b; Gouët and Leeuwis, unpublished). Interaction among farmers contributes to enlargement of the rural population's social capital (Uphof, 2000,

---

[102] The research project 'the Web of development' aims to contribute to a better understanding of the role of RPOs in development trajectories. The project is conducted at Wageningen University by Christian Gouët under the supervision of Prof. Cees Leeuwis (http://www.onderzoekinformatie.nl/en/oi/nod/onderzoek/OND1309727/).

[103] In our other chapter we will illustrate the development cooperation of Agriterra by one of the first methods deployed: the peer-to-peer exchanges, and advisory missions organised through AgriPool.

Berdegué, 2001, Blokland, 2003), contributes to a country's development and democracy building social networks and 'trust' (Paxton, 2002; Blokland, 2003) and generates the necessary conditions for innovation and technological diffusion, leading to technical and social change (Engel, 1995, 1997; Leeuwis, 2004).

In an effort to elucidate these questions, the main purpose of Gouët's current research is to examine whether the Agriterra axiom continues to apply in the changing circumstances of worldwide market liberalization; and if it indeed does, then to explore and identify bodies of theory relevant to understand the role of RPOs as development fostering engines. Likewise, the research could also explore implications of these theories for the practice of capacity building in RPOs.

The above comes at a point in time where there is growing interest in the role and potential of RPOs for development. The 2006 Arnhem conference on Farmers Fighting Poverty brought bilateral and multilateral donors together with representatives of NGOs and agri-agencies, all very interested in seeing RPOs play a major role. The World Development Report 2008 has Agriculture and Development as its main subject. A significant number of programmes and agencies have emerged that aim to stimulate the formation of RPOs and strengthen their capacity to contribute to development (Gouët *et al.*, unpublished). In the Netherlands this focus on RPOs and the collective learning on the subject has been organised within Agri-ProFocus, the public-private partnership for agricultural producer organisations. Growing interest and increasing support for these types of organisations in development cooperation calls for a more solid theoretical foundation regarding the relationship between development and the existence of RPOs; and for a more solid theoretical underpinning of the capacity strengthening efforts of development agencies. More empirical evidence as to impact is also needed. This means answering, both from a theoretical and practical perspective, the following question: do strengthened farmers' organisations indeed have a positive impact on development, democratization, income distribution, and reduction of poverty?

References

Agriterra, 2000. Annual Report 2000. Agriterra, Arnhem.
Agriterra, 2006. Farmers Fighting Poverty. Producers' Organisations Support Programme DGIS-Agriterra (POP) 2007-2010. Agriterra, Arnhem.

Berdegué, J., 2001. Cooperating to Compete. Associative Peasant Business Firms in Chile. PhD Thesis, Wageningen University, the Netherlands.

Biekart, K., 1999. The politics of Civil Society Building. International Books, Utrecht.

Biekart, K., 2000. PIPGA: Programa de Investigación Participativa Generadora de Alternativas de Desarrollo. Agristudies (Available through www.agriterra.org.). Agriterra, Arnhem.

Blokland, K., 1992. Participación Campesina en el Desarrollo Económico. La Unión Nacional de Agricultores y Ganaderos durante la revolución Sandinista. Ph.D. Thesis. PFS, Doetinchem.

Blokland, K., 1995. Peasant alliances and 'Concertation' with Society in Bulletin of Latin American Research, Vol. 14, nr. 2. Elsevier Science Ltd, London

Blokland, K., 1996. Neoliberalism and the Central American peasantry in: Liberalization in the Developing World by André Mommen and Alex Jilberto Fernandez, Routledge, London.

Blokland, K., 1999a. The decollectivization dilemma. Theoretical reflections on land use and tenure. Agristudies (Available through www.agriterra.org.). Agriterra, Arnhem.

Blokland, K., 1999b. The PIPGA case and implications for the strengthening of Producers' Organisations. Contribution to the IFAP-World Bank seminar on Strengthening of Producer Organisations. Agristudies (Available through www.agriterra.org.). Agriterra, Arnhem.

Blokland, K., 2002. Do producer organisations foster economic development? Agristudies (Available through www.agriterra.org.). Agriterra, Arnhem.

Blokland, K., 2003. From Plunder to Economic Democratisation: Towards a new development paradigm of farmers worldwide. Agristudies (Available through www.agriterra.org.). Agriterra, Arnhem.

Dart, J.J. and R.J. Davies, 2003. A dialogical story-based evaluation tool: the most significant change technique, American Journal of Evaluation 24, 137-155.

Diepenbeek, W. van, 1990. De Coöperatieve organisatie. Coöperatie als maatschappelijk en economisch verschijnsel. Eburon, Haarlem.

Engel, P., 1995. Facilitating innovation: an action-oriented approach and participatory methodology to improve innovative social practice in agriculture. PhD thesis, Wageningen University, the Netherlands.

Engel, P., 1997. The social organisation of innovation: a focus on stakeholder interaction. Royal Tropical Institute (KIT). Amsterdam, 1997.

Engel, P., 2002. Global Chains: chain gangs or development opportunities? Symposium proceedings of the 8th lustrum of Mercurius - Wageningen. Wageningen Academic Publishers. The Netherlands, 2002.Engel, P.

Gouët, C., 2004. Spinning a Web of development. The evaluation of Agriterra's support to IFAP (2001-2003). Agristudies (Available through www.agro-info. net).

Gouët, C. and C. Leeuwis, 2004a. Towards Capitalizing on Capacities. The evaluation of Agriterra's programme 2001-2003. Agristudies (available www. agro-info.net).

Gouët, C. and C. Leeuwis, 2004b. Capacity Building of Rural People's Organisations at the local, national and international spheres. Summary of the evaluation of Agriterra's programme on international cooperation between rural people's organisations (2001-2003). Wageningen University, report document.

Gouët, C. and C. Leeuwis. Theoretical perspectives on the role and significance of RPOs in development. Implications for capacity development.

Gouët, C., and A. van Paassen (Unpublished): RPOs and small farmers' market access in times of globalisation of agri-food market: the view of the actors involved.

ICA - International Cooperative Alliance, 1997. Co-operatives and the Poor. Co-operative College, Loughborough.

IFAP - International Federation of Agricultural Producers, 1995. Negotiating Linkages. Farmers organisations, agricultural research and extension. IFAP, Paris.

IFAP - International Federation of Agricultural Producers, 2005. Women Farmers' Leadership Program: for a stronger representation. Standing Committee On Women In Agriculture. Montevideo, Uruguay. November 7, 2005. IFAP Document WO 4/05.

Leeuwis, C., 2004. (with contributions by A. Van den Ban). Communication for rural innovation. Rethinking agricultural extension. Blackwell Science, Oxford.

Morré de, D., 2002. Cooperación campesina en los Andes. Un estudio sobre estrategias de organisación para el desarrollo rural en Bolivia. Ph.D. Thesis, Koninklijk Nederlands Aardrijkskundig Genootschap/ Faculteit Ruimtelijke Wetenschappen Universiteit Utrecht. Utrecht 2002.

Paxton, P., 2002. Social Capital & democracy: An interdependent relationship. American Sociological Review 67. ASA, Washington.

Presno, N., 2004. Gestión social como estrategia competitiva de las cooperativas agrarias. Specially prepared for the IFAP-DCC seminar on capacity building needs for smallholders' cooperatives. Uruguay 2004.

Quiroz, R., G. Quiroga and C. Gouët, 2005. Cooperativas y empresas campesinas asociativas para el desarrollo de iniciativas de mercado (Cooperatives and small holders' associative companies for market initiatives). Report after an international seminar and workshop hold in Montevideo, Uruguay. October 2004. Agristudies (Available through www.agriterra.org.).

Rondot, P. and M. Collion, 2001. Agricultural Producer Organisations; their contribution to rural capacity building and poverty reduction. Summary of a workshop, June 28-30 1999. Rural Development Department, World Bank, Washington DC.

Schiff, M. and A. Valdéz, 1992. Plundering of Agriculture in developing countries. World Bank, Washington.

Schuurman, J., 2006. Farmers Fighting Poverty Brochure. Agriterra, Arnhem.

Spoor, M., 2002. Policy regimes and performance of the agricultural sector in Latin America and the Caribbean during the last three decades. Journal of Agrarian Change Vol. 2, no. 3.

Stessens, J. and C. Gouët, 2004. Efficient contract farming through strong Farmers' Organisations in a partnership with Agribusiness. Hoger Instituut voor de arbeid, Katholieke Universiteit Leuven. July 2004.

Stockbridge, M., A. Dorward and J. Kydd, 2003. Farmer organisations for market access: learning from success. Briefing paper, downloaded 17 December 2006 from www.cphp.uk.com/uploads/disseminations/R8275%20040516%20Bfg%20Paper%20FO%20for%20market%20access.pdf.

Uphof, N., 2000. Understanding Social Capital: learning from the Analysis and Experience of Participation. Published in: Social Capital. A Multifaceted Perspective (Dasgaputa, P. and Serageldin, I. Eds. (2000). World Bank Publications.

Uphof, N. and C. Wijayaratna, 2000. Demonstrated Benefits From Social Capital: The Productivity of farmer Organisations in Gal Oya, Sri Lanka. World development, 28:11 (November 2000).

Zamosc, L., 1987. La cuestión agraria y el movimiento campesino en Colombia. Luchas de la Asociación Nacional de Usuarios Campesinos (ANUC), 1967-1981. UNRISD, Genève/CINEP, Bogotá.

The rise of new rural producer organisations in China

Jos Bijman, Rik Delnoye and Giel Ton

The Chinese government has developed a policy to strengthen rural development through strengthening the competitiveness of the agricultural sector. Producer organisations are seen as important elements to realize the economic potential of the agricultural sector. Therefore, the national government has put in place a new law targeting cooperatives, which will be effective as of July 2007. This chapter also presents several case studies that addresses the economic, political and organisational challenges currently facing POs.

1. Introduction

The economy of China is booming with annual growth of 8-12 percent over the last decade. However, development and growth is mostly concentrated in the Eastern part of China, and mainly in urban areas. The rural areas, and particularly those in the Western parts of China, are lagging in terms of economic development. The income ratio between rural and urban areas is approximately 1 to 3, and is still increasing (Feng, 2006). This is not to say that rural China has not yet experienced any changes. Already major achievements have been obtained: productivity has grown rapidly, exports are rising, and rural income has risen dramatically. Still, one of the major challenges for Chinese society in general is to make the rural economy an integral part of modernization.

A booming economy and political liberalization provide opportunities for farmers to shift to new lines of production, enter new markets, and set up new supply chain relationships. In order to strengthen their bargaining position, to exchange experiences, and to facilitate innovation, farmers are establishing producer organisations (POs). In recent years, the number of POs is rapidly increasing, although the absolute number of POs is still relatively low (Shen *et al.*, 2005).

One of the interesting issues in studying POs in China is the role of the government in supporting these organisations. While many POs in China, as in other parts of the world, are voluntary member-based organisations, state influence is still very strong. New legislation on cooperatives, which will be effective from July 1, 2007, emphasizes that POs are farmer self-help organisations, established for the benefit of members and run by members. However, the law also encourages state agencies at regional and local level to actively support the establishment of new POs. Finding a balance between state support on the one hand and maintaining independent organisations on the other, is one of the major challenges for new POs in China.

This chapter presents background information on the rise of new rural POs in China and it discusses several of the challenges these POs face. It starts with a brief introduction to the recent political and economic changes, which have tremendous impact on rural development. The new law on cooperatives is an important change in the institutional environment for POs. The next section discusses the role of POs in the changing agricultural markets where market liberalization and the rise of supermarkets make it more difficult for small farmers to compete. Subsequently some figures are provided regarding new producer organisations in China. A number of case studies present more practical information on the challenges of these POs. We conclude with a number of observations on the challenges for organising effective farmer-led associations and cooperatives.

2. Economic and political changes in China

Since the early 1980s, China has been restructuring its economy from a planned economy towards a market-oriented economy. This transformation process has already led to fundamental changes in the economic landscape of China. For instance, in the 1990s, per-capita grain output reached a level similar to that of developed countries (Sonntag *et al.*, 2005). Rising food exports demonstrate that China's farmers can now compete in international markets. Rural incomes have risen significantly and hundreds of millions of people have escaped poverty.

However, many rural areas do not (fully) participate in this economic modernization. China's 200 million small-scale farming households still follow traditional production and marketing patterns. However, market developments dictate a need for increasing the scale of production, enhancing product quality, and improving the efficiency of supply chains. The challenge for Chinese policymakers is to promote agricultural modernization in such a way that rural areas, particularly those more distant from large urban developments in the East, will also benefit from the overall economic growth. In addition, the country as a whole is experiencing a shift from a rural economy towards an urban economy; from agriculture towards industry and services. Thus, the challenge for China is to establish effective linkages between rural and urban areas and to encourage a shift in labour from agriculture to other industries.

Chinese political leaders have acknowledged that policy reforms are needed to guide economic transformations. Since the mid-1990s, Chinese agricultural officials have promoted a 'companies plus rural household' strategy to bring farmers into the commercial food sector and raise incomes. This strategy emphasizes links between farmers and processing and marketing companies to strengthen farmers' connections with the market. So-called 'Dragon head' companies are selected or established by local government authorities to contract with farmers to procure produce with specific attributes. The dragon head company provides seed, operating loans, fertilizer and other inputs, and technical expertise. Local government continues to play a major role in coordinating these linkages between farmers, processing industry and markets.[104]

In 2004, the central government issued the 'Document No. 1', which exclusively deals with agricultural and rural development. This document specifically pointed out that farmers' professional associations should be encouraged. The government regards the establishment of POs as essential for strengthening the competitiveness and in realizing the economic potential of the agricultural sector. Enhancing competitiveness is necessary as China has agreed to open up its markets as part of its obligations under WTO membership. In addition, to be able to continue the combination of high economic growth and low inflation China requires that food prices stay low and that the shift of low cost labour to other industries continues.

2.1 A new law on cooperatives

On October 31, 2006, the 10th National People's Congress of the P.R. of China adopted legislation supporting professional farmer cooperatives. The law was later ratified as Order 57 by President Hu Jintao. Presidential Order 57 will become effective on July 1, 2007. Article 1 stipulates the reasons for developing this special law: 'Its purpose is to facilitate and direct the development of farmer Co-operatives, standardize organisation and behaviours of them, protect legal interests of Co-operatives and members, and foster growth of agricultural and rural economy.' Thus, this national law on professional farmer cooperatives provides the legal framework for the establishment and functioning of farmer cooperatives.

A law on cooperatives is of great importance for farmers and their economic organisations. So far, much of the group action by farmers has been informal, or it has fallen under various, often local, and potentially conflicting, regulations. A national law, applied by provincial and regional authorities provides a uniform legal framework for the whole country.

---

[104] There are some doubts as to the viability of this approach (World Bank, 2006; Gale, 2003). Local government officials may not have a sufficient knowledge base or the financial capacity necessary to guarantee success. In addition, the ownership structure of the dragon head companies is not clear. Many seem to be spin-offs of local grain bureaus and other government marketing entities. Management decisions seem to reflect government plans to develop particular sectors, sometimes resulting in overproduction. Finally, small-scale farmers may not benefit much from this approach, as they lack bargaining power when signing contracts with the dragon head companies.

The importance of legislation for farmer cooperatives has been emphasized by many cooperative scholars. For instance, Münkner (2005) states that 'legislation can make it easier to form and run cooperatives, if it offers a clear and easily applicable model, helping the organisers to set up the rules necessary for operating joint/group activities in such a way that typical conflicts (which may arise among members or among members and the organisation) are as much as possible avoided.' Without legal status the organisation lacks access to credit, cannot enter into legally binding contracts, and there is no legal standing for the mutual obligations and responsibilities between cooperatives and their members.

While enacting this new law on cooperatives is a very important step towards helping farmers to improve market access, capital and knowledge, there has been some concern about the emphasis on cooperatives as a tool for realising governmental policies for rural development. Münkner (2006: 11) states that it is an old mistake to 'use co-operatives as instruments for implementation of government policy, rather than perceiving co-operatives as voluntary and independent self-help organisations of their members. Where co-operatives work as development tools in the hand of government, it is difficult to see how co-operatives can avoid becoming dependent on government.' Münkner argues that state sponsorship usually turns into state control, which is not only against cooperative principles, but is even counterproductive as state-supported cooperatives usually fail to generate member commitment (which in turn is needed for collecting capital, effective decision-making and efficient use of cooperative facilities).

Fock and Zhao (2006) also call for caution in direct government intervention in the formation and operation of farmer groups. In addition, they are concerned about membership heterogeneity as the law allows for firms and government-affiliated agencies to become members of farmer cooperatives. Heterogeneity of members often causes conflicts of interests within the group, which thereby lead to inefficient and ineffective decision-making (Bijman, 2005). The rationale for allowing non-farmers to become members of a cooperative lies in the common interest for building integrated supply chains or modernizing agriculture. However, it poses serious risks for member commitment (and thereby for the sustainability and efficiency of the cooperative) as local state agencies and leading enterprises can dominate. Fock and Zhao argue that it would be better to develop separate programs aiming to improve supply chains efficiency and to promote the restructuring of agriculture.

3. Changes in agricultural markets

Markets for agricultural inputs and outputs are rapidly changing in China. During the period of the planned economy, agricultural inputs and outputs were subject to a system of unified supply and marketing. Since the start of reforms in the early 1980s, markets for agricultural inputs have gradually opened up, although in some parts of the country government agencies are still the main source of seeds, fertilizers and agrochemicals under a system of government licensing and state trading. Most agricultural markets, however, are now open to competition.

At least three major developments in agrifood markets are pushing for structural change in the agricultural sector, particularly affecting small producers (World Bank, 2006: 13). First, China's entry into the WTO has altered the dynamics of domestic markets. WTO membership has created opportunities in areas where China has comparative advantage, including livestock, aquatic and horticultural products. However, exports are only possible if high and stringent quality standards can be met (Huang and Gale, 2006). Many small-scale farmers using traditional production methods have difficulty in meeting international standards and in providing the necessary certification to guarantee that the standards are met.

Second, increased consumer sophistication means that they are no longer content with a limited choice of products or seasonal availability and have a growing awareness of food safety issues. However, small producers may have a hard time complying with these quality and safety requirements. Demand for uniform products of large quantities as well as supply chain measures such as cold-chain transportation, all favour large-scale production over small holder farming. Food safety concerns have led to the creation of a variety of certification standards, but the cost and effort involved in compliance to these standards is often beyond the means of individual small scale farmers.

Third, the rise of supermarkets as major food outlets is resulting in supply chain restructuring, which may make it more difficult for small farmers to compete. Let us take a more detailed look at the rapid expansion of supermarkets in China and its effects on supply chain organisation.

Traditionally, wet markets have been the main place for rural and urban consumers to purchase fresh agricultural products. Nowadays, supermarkets are taking over this role, at least in urban areas. Government policy has not been favourable towards wet markets, pressuring them to convert into modern supermarkets in a move to control hygiene and to reduce congestion.

While modern supermarkets, convenience stores, and hypermarkets were nearly non-existent in the early 1990s, they have now captured an estimated 30 percent of the urban food market and are growing at rates of 30 to 40 percent annually (Hu *et al.*, 2004). Supermarkets first developed on the south and east coasts of China, and then in major northern cities like Beijing, Tianjin and Dalian. More recently, supermarkets have been penetrating into western inland China. Consumers targeted by supermarkets were at first the rich and middle income segment of the population in big cities, however, this is now expanding to include lower income segments of the population and smaller cities.

Supermarkets in China were initially small-scale shops that focused on food, especially processed food. When foreign companies, such as Tesco and Carrefour came on the scene, fresh produce also became an important category for supermarkets. This foreign direct investment stimulated the development of domestic supermarket chains, and the dominating types of stores in the supermarket sector are now large scale hypermarkets. Moreover, former state-owned food stores privatized, went into bankruptcy or were taken over by modern supermarket chains. Domestic supermarket chains advanced aggressively to inland and secondary cities, aiming to firmly establish themselves in the market before foreign companies entered those regions. It is estimated that in big cities about 30% of all fresh produce is retailed through supermarkets.

The growth of supermarkets in China can be related to an increased demand for the specific services they provide: a result of rapid urbanization; per capita income growth; the increasing employment of women; the 'westernization' of life styles, particularly amongst the younger generation; the reduction of effective food prices for consumers due to supermarkets' greater ability to control costs through economies of scale; improved logistics; growing access to refrigerators, which allows larger quantities of food to be stored; increased access and use of cars, allowing shopping to be done away further away from home.

The rise of supermarkets in retailing (fresh) agricultural products has major implications for the supply chain. Supermarkets in general favour centralized procurement systems, specialized and dedicated wholesalers, preferred supplier systems, and private standards for fresh produce (Shepherd, 2005). The purchasing practices of supermarket chains, with their demand for standardization and high quality products, large volume purchases, as well as their practice of delayed payments, mitigate in favour of large-scale production by well-resourced suppliers. Small producers find themselves barred from access to the high-value market and trapped behind an insurmountable wall that separates them from potential consumers.

These developments in the agricultural markets and food retail business are making it harder for small farmers to access markets and to compete. It is exactly for this reason that producer organisations are so important, as they can be (a part of) the solution. By establishing POs, small farmers have an instrument to face the challenges that come with restructuring agrifood markets.

4. New producer organisations in China

Voluntary POs are not a completely new phenomenon in China. In the first half of the 20th century, before the founding of the communist state, some parts of China had experience with voluntary producer-led cooperatives (Du, 2002; mentioned in Hu *et al.*, 2007). However, from the 1950s onwards, cooperatives were used by the government to centrally control and manage agricultural production, trade and consumption. Under the system of collective farming, the supply of inputs, agricultural production itself, and distribution of the products were all directed by the central plan. All collection and distribution of farm products was managed by state agencies under the 'Unified Procurement and Sales System (UPSS)'. In the case of vegetables and pork, so-called category II products, supply and sales cooperatives procured farm products, while state-owned trading companies retailed them to urban consumers. This situation continued until the market liberalizations of the mid 1980s.

In the 1980s, collective farming was gradually transformed into family-based farming, while the unified collection and distribution system was substituted by a more market-oriented system. Since the abolishment of the UPSS in 1985, markets have fully liberalized for fruits and vegetables, pork, seafood, eggs, and other agricultural products (with the exclusion of cotton and grains). Different marketing channels developed for these products, with farmers, government agencies, traditional cooperatives, state farms, and private traders all become marketing enterprises. Wet markets, which were under restriction during the planned economy, came back to life soon after the reforms. In addition, wholesale markets and professional traders rapidly expanded their business, as interregional trade was fully liberalized. Within these liberalized domestic markets, POs can play an important role in linking producers with customers.

Producer Organisations are known under different names in China: farmer associations, farmer professional associations, farmer professional cooperatives, farmers' cooperative associations, farmer cooperative economic organisations, etc. One distinction often made is between associations and cooperatives. Although the term association (*xie hui*) is not very precise, it is used in World Bank publications on POs in China as well as by Sonntag *et al.* (2005) in their extensive review of China's agricultural and rural developments. This term started to appear in the 1980s, mostly referring to farmer organisations established to provide members with technical support and information. Associations can be informal as well as formal. While 'association' covers virtually every kind of farmer organisation, the term 'cooperative' (*hezuo she*) is more restricted, but equally ambiguous. For farmers the word cooperative generally has negative connotations as the rural areas of China only started to prosper with the dismantling of communes and the introduction of the family responsibility system, which marked the end of the monopolistic position of government owned commercial organisations operating as 'cooperatives'. Nevertheless, the term 'cooperative' is beginning to dominate parts of the discussion on farmer associations that carry out economic functions. The term is used in the 2005 Zhejiang Province law on farmer professional cooperatives (*nongmin zhuanye hezuoshe*) as well as in the new national law on farmer professional cooperatives. Finally, translation of the Chinese terms into English adds to the confusion. For instance, the term *zhuanye* can be interpreted either as 'professional' or as 'specialized'. While in World Bank publications the term professional is used (as in farmer professional association), Hu *et al.* (2007) use the term specialized (as in farmer specialized association and farmer specialized cooperative).[105]

According to Hu *et al.* (2007) associations and cooperatives differ in terms of the ownership of fixed assets, as well is in their main function. The farmer professional cooperatives are similar to cooperatives in

---

[105] In this chapter we adopt the World Bank custom of translating *zhuanye* into professional.

Western Europe and North America: they are engaged in the marketing and processing of agricultural products, and they own fixed assets. Farmer professional associations (FPAs) are mostly set up to provide technical assistance and to share information amongst members. They do no charge a membership fee and do generally not own fixed assets. The term FPA is rather broad, and includes both organisations with thousands of members and community organisations with only a dozen farmers. Professional cooperatives can be found mostly in the Eastern (and more developed) parts of China, while the FPA dominates in Western and Central China (Hu *et al.*, 2007).

Detailed figures on the number of POs in China are hard to come by. On the basis of a nationwide survey in 2003, Shen *et al.* (2005) conclude that about seven percent of all villages in China had functioning FPAs and about two percent of farm households in China were members of such FPAs. Hu *et al.* (2007: 442) report data from the Rural Development Institute of the Chinese Academy of Social Sciences, stating that in 2004 China had more than 150,000 POs, of which 65% were FPAs and 35% were farmer professional cooperatives. Data on the number of POs in China must be interpreted with care, due to the above mentioned difficulties of names as well as because many POs are not formally registered (and, some of the formally registered POs are not functioning).

Even taking into account these problems in accessing statistics, it is fair to say that the development of POs in China has not been substantial. Particularly addressing cooperatives, Zhou (2004) gives three reasons why farmers may be reluctant to cooperate in POs. First, farmers may still have bitter memories from past experiences with cooperatives and the people's commune system (pre-1980s). Second, until recently the government has been rather reluctant to promote cooperatives and other POs. Third, Chinese farmers are experimenting with a wide range of collaborative arrangements, of which the cooperative is only one.

5. Case studies

We now present three brief case studies of POs in three different Chinese provinces. All three cases provide information as to the challenges that newly established POs are facing in their internal organisation, in their relationship with state agencies, and in their quest for capital. The first case addresses cooperatives in Zhejiang Province, on the east coast just south of Shanghai, and is particularly interesting because Zhejiang has had a provincial law on cooperatives since 2005. One of the challenges for POs in Zhejiang province is how to balance member interests with shareholder interests. A second case looks at POs in Yunnan, a province in the south of China, bordering Burma and Laos. This mountainous province is one of the poorest regions in China. The challenge presented in this case is how to combine the interests of government agencies supporting (the establishment of) POs and of the members to maintain full control over the PO. The third case concerns POs in Hunan Province, in the central/southern part of the country. Hunan is a major fruit producing region, with a good climate for many fresh produce crops. However, it is not close to major urban areas, so supply chain management is of crucial importance to farmers in this region. Federated cooperative structures can perform these joint supply chain functions. One of the main challenges for POs in Hunan is to find a balance between marketing activities at the local and regional levels.

5.1 Zhejiang Province[106]

Zhejiang Province is located on the east coast of China, south of Shanghai. The province has about 46 million inhabitants. For the development of POs in China, Zhejiang Province is of great interest; it was chosen by the national Ministry of Agriculture as a pilot for promoting the establishment of farmer professional cooperatives. In 2004, Zhejiang Province had 1789 farmer professional cooperatives and 1019 farmer professional associations. In 2005, the provincial government passed a provincial law on farmer professional cooperatives.

The main reasons for farmers in Zhejiang Province to become members of a cooperative are to access markets and to obtain technological support to improve on-farm productivity and quality. This demand also influences farmer selection of board members. They are more likely to select as members of the board of directors, those farmers that have large farms themselves, as these have proven to be successful in terms of market access.

One of the interesting characteristics of farmer professional cooperatives is the relationship between shareholders and members. In Zhejiang Province cooperatives, not all members are necessarily shareholders. This means that the equity capital of the cooperative is provided by a subset of the membership. In addition, not all shareholders

---

[106] This description of the development of cooperatives in Zhejiang Province is based on an empirical study carried out by Hu *et al.* (2007).

are members. This means that the cooperative may obtain equity capital from persons, firms or other organisations that do not use services of the cooperative. In Zhejiang Province, one can find cooperatives where all members are also shareholders as well as cooperatives where only one percent of members are shareholders. On the basis of shareholder/member ratio, the cooperatives in Zhejiang Province can be divided into two groups. In one group only a minority of members are shareholders. The membership of the cooperatives in this group consists of a small number of core members (being large shareholders) and a large number of common members (who are just users). In the other group of cooperatives a majority of the members are also shareholders. But even in this group, shareholding is quite concentrated, and the big shareholders play a dominant role in providing equity capital.

In general, large farmers play a major role in Chinese cooperatives, as shareholders, as members of the board of directors, and as major users of the services of the cooperative. With the success of their individual farm, their relationships, their knowledge of markets, and their access to technology, these farmers/directors are considered the right people to lead the cooperative. In addition, the relationships that core members have with common members, for instance through kinship, signal the commitment of the core members to the cooperative. The director's ability to access both technology and downstream markets is a reason for common members to grant substantial decision-taking rights to the director(s). Hu *et al.* (2007) emphasize that farmer cooperatives in China are rooted in the Chinese traditional culture of personal relationships. Thus, efficient and effective governance is not only obtained by formal structures, but also by social mechanisms such as social control, commitment and trust.[107]

5.2 Yunnan Province[108]

In Yunnan Province, especially in the hilly border regions, farmers are resource-poor, having access to, on average, not more than half a hectare of land. Infrastructure is, however, improving rapidly, enabling access to markets and opening up new economic possibilities. In order to capitalize on new these opportunities collective action in organising

---

[107] Personal relationships can support the efficiency and effectiveness of the cooperative, as Hu *et al.* (2007) argue, but may also lead to oligarchy and nepotism.
[108] Information on POs in Yunnan Province has been collected mainly through interviews with PO leaders and government officials, by Rik Delnoye, in 2006.

supply and marketing is crucial. Associations and cooperatives are formed around major commercial commodities such as dairy, fruits, vegetables, and coffee or new promising cash crops such as macadamia nuts, peppers, and oranges. A typical PO has 100 to 200 members. Establishment of POs is often initiated by local governments within broader support programs promoting the production and marketing of selected crops. In addition to regional economic targets (e.g. reaching particular quantity levels in production) such programs have pro-poor objectives such as supporting farmer income.

We now turn to the example of a PO from Yinjiang County in DeHong Prefecture. The county administration is actively promoting the establishment of farmers associations. One of the first associations established in the county, in 2002, was a potato association. Potatoes are a new crop in the region. The production and marketing potentials of potato are promising. So far, the association has been very successful in marketing. Customers such as traders and processors are responding positively to the possibility of contracting large volumes at once rather than having to deal with a large number of individual farmers.

The potato association was established in the village by the administrative authorities. From the very start, the potato association received major support from the local government, such as a founding grant, office building and technical support. In 2006, the association had 244 members and 13 personnel. The personnel are remunerated by the local government but the objective is that the association will be able to pay its own personnel in the near future. In 2005 the association signed a contract for delivering 4800 tons to a processing company. In 2006, contracts for delivering 15 000 tons were signed, for which 3200 households were contracted, totalling an area of 13 000 mu (= 870 hectares).

The potato association applies a cooperative model with members and (separate) shareholders (similar to cooperatives in Zhejiang Province). For its members the cooperative functions as a service provider while for the shareholders receiving part of the profit of the cooperative is the main benefit. Shareholders do not hold voting rights. Of the profit, 30% is paid to shareholders.

Farmers and their newly established PO find themselves in an ambiguous situation. They still lack the capacities to manage the organisation themselves and lack the financial means to recruit professional assistance, while at the same time, the local government is eager to provide support. But instead of strengthening the human capacity amongst members in order to create a pool of potential leaders, government agencies often place their own staff in key functions (e.g. financial manager) and thereby undermine the independence of the PO. This threatens to reduce member commitment to actively participate, to comply with cooperative regulations, and to provide equity capital. It may even reduce the willingness of farmers to join the cooperative (depending on other alternatives available).

At present the leaders of the association face several challenges, both on operational and organisational issues. As hundreds of farmers have indicated that they are interested in becoming members, the cooperative has to scale up its operations and thus needs investment capital. In addition, the leadership is looking for equitable ways to remunerate the efforts of members who first joined the cooperative. How can shareholders be satisfied (and even new shareholders added) without making profit growth into the main objective of the cooperative? In addition, leaders are looking for the optimal decision-making structure. Most of the current members are living in the village where the association was established but the organisation plans to expand to neighbouring townships. Presently the board of the association is comprised of elected leaders (mainly the initiators of the association). The future governance structure, particularly the representation of new townships, is still under discussion. The present leadership fears an increase in membership heterogeneity but at the same time sees strong potential in the production capacity new members would bring to the cooperative.

From this brief description of new POs in Yunnan Province we can draw the following conclusions. Market developments are fostering the establishment of POs. By joining resources, farmers can strengthen their bargaining position, can develop supply chain linkages, and can obtain access to technical and marketing expertise. However, the major challenge is to set up and maintain an independent farmer-led organisation, while obtaining some support from governmental agencies without letting the government control the PO. Farmers and farmer representatives are well aware of this challenge, and time and again emphasize the need for establishing truly self-help, farmer-led organisations rather than organisations that are development tools in the hands of government.

5.3 Hunan Province[109]

Hunan is a province with a large number of POs, both in absolute and relative terms (Zhou, 2004). In 2002, Hunan counted almost 15000 farmer professional cooperatives (the second highest number of all provinces in China). Hunan is a major fruit producing province. We visited several orange producing cooperatives in Fenghuang County and in the city of Zhangjiajie, the main city in the County. These cooperatives had recently been formed and were in the process of establishing a federative structure to carry out joint marketing activities.

The South Great Wall Green Fruit Industry Cooperative brings together the associations of several villages around the town of Liaojiaqiao. It is a cooperative consisting of 500 members coming from 20 villages. The membership is rather heterogeneous, with fruit growing families, fruit processing families and fruit selling families. Current activities are concentrated on quality improvement and brand name development assisted by the local office of the Agricultural Bureau. The cooperative has entered into several interesting contracts with middlemen, however it is not able to sell all of the products of its members. The marketing activities of the town cooperative are in addition to those of the village associations and individual members. The town cooperative is commissioned to market the products of local village associations, but sometimes finds it difficult to deliver the right quantity and quality because the village associations and individual members can also make deals on their own at the same time.

The financial sustainability of the cooperative is still weak. No commission fee is charged, and current costs of the cooperative are assumed by the (richer) board members as being part of their individual marketing costs. Training and technical support is provided free-of-charge by the Township Government. As no significant internal income is generated and costs are assumed rather informally by the board members, priority in this cooperative seems to be the strengthening of the internal organisation and clarification of the benefits for the township cooperative as compared to the village associations. As many members are also involved in pig raising and horticulture, internal policy discussions take place about whether to maintain the focus on fruits or diversifying into other members' products.

---

[109] The information on POs in Hunan Province was collected, mainly through field interviews with PO leaders and government officials, by Giel Ton, in 2005.

Fruit cooperatives from different towns and villages have formed a regional cooperative on the county level. The membership of the regional cooperatives consists of three township level cooperatives and thirteen village level cooperatives. The regional cooperative started operations in September 2004. Membership fees vary by size of the member: 50 yuan for big cooperatives and 30 yuan for small cooperatives. The regional cooperative represents 5000 fruit growers. The main activity is to sort and market oranges. In 2004 the cooperative exported to Canada, organised a wholesale market, and participated in business fairs. The county cooperative manages to sell approximately 30 percent of the total produced y members. Additional member services include providing market information and indicating the quality requirements of potential buyers. One of the challenges for the regional cooperative is to maintain the trust of its members, particularly on the issue of price determination. Board members receive training in marketing, but pure commercial activities are delegated to professional staff. Another challenge is to maintain a clear division of tasks between the regional, town, and village cooperatives.

6. Discussion

The rural economy in China is experiencing major transformations. While most farmers are still small scale producers, they are pressured to adapt to rapidly changing agricultural markets. Producer organisations are a major tool in the hands of farmers to improve access to and position in markets and thereby raise incomes. In setting up and running associations and cooperatives, farmers are faced with a number of organisational and operational challenges. Governments are both part of the solution and part of the cause. In this concluding section we will summarize the major challenges for Chinese POs as they have appeared throughout the chapter.

First, economic and political reforms provide many opportunities for farmers and thereby for the establishment of new POs. However, the new national law on cooperatives is sending ambivalent signals to farmers and local government officials. On the one hand it emphasizes that POs are self-help organisations for farmers who associate voluntarily and maintain control over their organisation. On the other hand it considers POs as organisations that can be entrusted with the task of implementing rural and agricultural development projects, which may turn cooperatives into agents for the implementation of government policy. Farmers planning to set up a PO, or those on the boards of existing POs, may find it hard to keep sufficient distance from state authorities, as they need financial and other support to get the PO functioning properly. Here, third party support from either within or outside of China may be helpful in training both farmers and government officials about the need to keep a clear separation between private and public roles.

Second, POs in China seem to be very heterogeneous in terms of membership. It is not uncommon for membership to include individual farmers, representatives from processing and trading enterprises, government agencies, and people from research institutes. Non-farmers often take the lead in managing the PO. While this may be rational from the perspective of getting different stakeholders to support the establishment of new POs, it usually leads to difficult decision-making processes. In addition, interests other than those of the farmers may become dominant, which may affect member commitment and thereby organisational efficiency. Related to the issue of heterogeneous membership is the distribution of income rights and decision-making rights. Having non-members as providers of equity capital may lead to conflicts of interest between members and investors. Members want favourable patronage conditions; investors want a high return-on-investment. Even when investors have no decision-making rights (no votes), the cooperative still has to take their interests into account otherwise they may withdraw investment.

Third, most small farmers do not have the knowledge and experience to actively participate in PO decision making, even though this is a basic conditions for the good functioning of a farmer-controlled PO. While in most developed countries generations of farmers have had experience with jointly owned, democratically run POs, farmers in China have only recently started to learn about the advantages and disadvantages of a PO as well as about their rights and obligations as members.

Fourth, the experience of PO in other countries has shown that, even when a favourable legal and regulatory framework exists, an independent catalyst is often needed to get POs started, for growth and expansion and for improvements in performance. In the Chinese context, such PO promoting agencies are preferably not state agencies, as it is important that the catalysts are first and foremost responsive to the needs of farmers and their organisations. Such catalysts could be cooperative extension services linked to (agricultural) universities, or they could be special independent institutes that promote POs. Such institutes would preferably be set up by the POs themselves. The main role of such institutes would be to provide training to PO leaders, to help lobby governments, and to facilitate the actual establishment of POs.

References

Bijman, J., 2005. Cooperatives and heterogeneous membership: eight propositions for improving organizational efficiency, Paper presented at the EMNet-Conference Budapest, Hungary, September 15 - 17, 2005.

Feng, S., 2006. Land rental market and off-farm employment: rural households in Jiangxi Province, P.R. China, Wageningen: Wageningen University (PhD Thesis).

Fock, A. and J. Zhao, 2006. Farmer-controlled organisations in China: pushed forward or taking-off? An assessment against international experience. International Conference on Promoting the Development of Farmer Cooperative Economic Organisation in China. Beijing (December 16-17).

Gale, F.H., 2003. China's growing affluence: how food markets are responding. Amber Waves, June 2003, Washington: USDA/ERS, (www.ers.usda.gov/ amberwaves/june03/features/chinasgrowingaffluence.htm)

Huang, S. and F. Gale, 2006. China's Rising Fruit and Vegetables Exports Challenge U.S. Industries. Electronic Outlook Report from the Economic Research Service. Washington, USDA/ERS.

Hu, D., T. Reardon, S. Rozelle, P. Timmer and H. Wang, 2004. The emergence of supermarkets with Chinese characteristics: challenges and opportunities for China's agricultural development. Development Policy Review 22(5): 557-586.

Hu, Y., Z. Huang, G. Hendrikse and X. Xuchu, 2007. Organisation and strategy of farmer specialized cooperatives in China. In: G. Cliguet, G. Hendrikse, M. Tuunanen and J. Windsperger (eds.). Economics of Management and Networks. Franchising, Strategic Alliances, and Cooperatives. Heidelberg, Physica-Verlag / Springer, pp. 437-462.

Münkner, H.-H., 2005. Answer to Questionnaire. World Bank Office in Beijing: International Workshop on Farmer's Cooperative Law Establishment, Beijing, 20-21 April.

Münkner, H.-H., 2006. Comments on the Farmer Professional Co-operatives Law, 2006, passed by the 24th Session of the 10th National People's Congress on October 31, 2006, University of Marburg, Germany.

Shen, M., S. Rozelle and L. Zhang, 2005. Farmer Professional Associations in Rural China: State dominated or new state-society partnerships? In: B.H. Sonntag, J. Huang, S. Rozelle and J.H. Skerritt (eds.). China's agricultural and rural development in the early 21st century. Canberra: Australian Centre for International Agricultural Research. ACIAR Monograph No. 116: pp. 197-228.

Shepherd, A.W., 2005. The implications of supermarket development for horticultural farmers and traditional marketing systems in Asia (revised paper). FAO/AFMA/FAMA Regional Workshop on the Growth of Supermarkets as Retailers of Fresh Produce. Kuala Lumpur (October, 4-7, 2004).

Sonntag, B.H., J. Huang, S. Rozelle and J.H. Skerritt (eds.), 2005. China's agricultural and rural development in the early 21st century. Canberra: Australian Centre for International Agricultural Research (ACIAR Monograph No. 116).

World Bank, 2006. China - Farmers Professional Associations; Review and Policy Recommendations (www.worldbank.org.cn/english/content/fpa_en.pdf).

Zhou, Z.-Y., 2004. China's experience with agricultural cooperatives in the era of economic reform. In: R. Trewin (ed.). Cooperatives: Issues and trends in developing countries. Canberra, Australian Centre for International Agricultural Research. ACIAR Technical Report No. 53: pp. 9-21.

Farmers' organisations in agricultural research and development: governance issues in two competitive funding programs in Bolivia

Giel Ton

This chapter analyzes the increasing involvement of farmer organisations in agricultural R&D support. It addresses the growing importance of competitive funds for agricultural R&D and the expanding role of farmers' federations as mediators between grassroots organisations and private service providers in the governance of the contracting process and contract conditions. Two R&D programs in Bolivia are examined to illustrate the point.

1. Introduction

In this chapter[110] we analyze the increasing involvement of farmer organisations in agricultural research and development (R&D) support[111]. In many countries, the financing and quality control of agricultural R&D systems have changed quite dramatically over the past few decades. These countries used to have R&D systems that relied almost exclusively on the public financing of NARIs (National Agricultural Research Institutes) and public sector extension services. Now, new forms of R&D systems have been introduced with competitive funding that is open to both NARIs and private service providers. Especially in value chain development, the boundaries between R&D and extension are increasingly diffuse and increasingly intertwined. In those new systems, farmer groups are considered to be the clients of R&D activities that are supplied by public and private service providers. Sometimes farmers' organisations are even involved in priority setting and governance of these R&D systems. Farmers' organisations involved in agricultural value chains and marketing are of special interest as they are knowledgeable about chain bottlenecks towards which R&D can be targeted. However, experiences with farmers' organisation involvement in R&D have been mixed. In some systems their involvement has resulted in a well-functioning R&D system, while in other systems the results have been rather disappointing (Echeverría *et al.*, 1996; Hussein, 2001; Alex and Rivera, 2004).

In this chapter we address the growing importance of competitive funds to finance R&D activities for farmer groups and the growing role of farmers' federations as mediators between grassroots organisations and private service providers in governing the contracting process and contract conditions. We illustrate this using case studies of two R&D programs in Bolivia.

---

[110] This is an adapted version a paper written on coffee R&D, with financial support of the Dutch Ministry of Agriculture - LNV (Ton and Jansen, 2007).
[111] We will use the abbreviation R&D (Research and Development) as a container concept that includes all research and extension activities related to solving bottlenecks for value chain performance.

2. Modern R&D systems: competitive grants for contracting service providers

Modern R&D systems with private research providers serving farmers' organisations are being introduced in various developing countries. The World Bank reviewed a range of different demand-led R&D systems (Alex and Rivera, 2004). The review suggested that emerging institutional arrangements are being widely tested and are transforming as part of an evolutionary process (Alex *et al.*, 2004). In spite of contextual differences, these new R&D arrangements tend to have a similar architecture:

- The R&D needs must be submitted by organised farmers.
- The selection of proposals is done by applying quality criteria and is increasingly related to the expected impact of R&D on value chain performance.
- A public call for proposals invites private service providers to elaborate a proposal to meet R&D needs.
- Proposals are evaluated by a specialized unit and one of the service providers is granted the contract.
- This service provider elaborates a detailed proposal in coordination with the farmers' organisation that originally submitted the R&D request.

This system gives farmers' organisations access to R&D support for (part of) their activities. Farmers' organisations that accept an R&D arrangement with a service provider will have to invest time and money in the process. In their decision making, they will assess the expected 'price' and 'quality' of the R&D services delivered and the expected benefits (for them). However, agricultural R&D investments are risk prone. Research and Development normally does not produce immediate results; practical solutions to the farmers' organisation may only emerge after a period of research and experimentation. So the risk of failure of R&D investments is always present. Farmers and farmers' organisations in developing countries are well aware of this risk, as they have much experience with support activities that fail to produce the promised results. They need credible and cost-effective procedures to adapt and fine tune the arrangements when expected outputs are not produced (c.f. Williamson, 2002),. Most contracts between farmers' organisations and private service providers allow for some adjustments when things do not proceed as expected. However, these contracts tend to follow standard formats, elaborated by the (inter)national designers of the R&D system. Even when adjustment mechanisms exist, the possibility for grassroots farmers' organisations to effectively use them is limited by high transaction costs. Questioning an R&D arrangement will cost time, money and, it may sometimes result in overt disagreement or conflict that could cost them their reputation with other support agencies in the future. Grassroots farmers' organisations will look for ways to prevent this negative image. That is why federated farmers' organisations (referred to here as farmers federations) are increasingly involved in R&D governance issues. They act on behalf of their member organisations and reduce the eventual costs of discussing R&D contract issues.

Farmers' federations can have divergent goals and activities. Some farmers' federations have a more or less homogeneous membership of grassroots organisations that represent farmers producing the same commodity. Other federations represent geographically based grassroots organisations which are comprised of farmers with very diverse activities who produce different commodities. Because of their goals and activities, organisations working with the same commodity will be more knowledgeable on problems and bottlenecks in the chain. They are crucial in the process of specifying and selecting the most

Box 1. Research linkages with Farmers' Federations.

A World Bank study in West and Central Africa (Bosc *et al.*, 2001; Hussein, 2001), on the linkages between research service providers and farmers' organisations concludes that strong, federated farmers' organisations tend to be a more effective mechanism for empowering farmers in technology development processes than, for example, simply using participatory methods or working with small farmer contact groups. Also Wennink and Heemskerk (2005) stress the importance of larger organisations like farmers' federations. This increases possibilities for the upward participation of different local groups of farmers and provides a mechanism for downward accountability to the voices of farmers in national R&D systems. The most successful organisations appear to be (Hussein, 2001):
- Possessing several levels - at least three - from grassroots to federation level
- Based on free membership around common interest
- Access to different sources of funding
- Based around successful and remunerative economic activities
- Benefiting external organisation support for animation, capacity building and input/marketing

important bottlenecks, and the R&D needs in the chain. In contrast, geographically based federations typically focus on issues relevant for the region or for agriculture as a sector, rather than being limited to a specific commodity. They tend to be more knowledgeable on issues related to general policy and legislation, like poverty reduction strategies, property rights, commercial law, and credit provision. In the design of an R&D system differences in roles between different types of farmers' federations must be correctly understood.

A proper distinction must be made between two different roles of farmers' organisations when they are involved in agricultural R&D systems:
- articulating research requests; and
- governing the institutional process and arrangements of R&D for smallholders.

Where grassroots organisations and commodity federations are especially knowledgeable on R&D requests for chain development, their higher level organisations tend to be more knowledgeable on issues around contracts, processes and the governance of R&D systems. Experiences in Bolivia (CIOEC, 2003; Muñoz, 2004; Ton and Bijman, 2006) suggest that there are typical strengths and weaknesses within each type of farmers' organisations in R&D systems (Table 1).

Table 1. Relative strengths and weaknesses of different types of farmers' organisations in their participation in R&D systems.

	Articulating R&D needs for chain development	Controlling the process and conditions for R&D for chain development
Grassroots organisations	+ +	–
Commodity based federations	+	+
Geographically based federations	–	+

3. Experiences with competitive funding arrangements in Bolivia

We illustrate the role of farmers' federations as process controllers using two R&D programs in Bolivia: SIBTA (Bolivian System of Innovation and Technology Transfer - Sistema Boliviana de Innovation y Transferencia Tecnológica) and PROSAT (Technical Support Services to Smallholder Producers - Servicios de Asistencia Técnica para Pequeños Productores). The farmers' federation that is the protagonist in these cases is CIOEC-Bolivia, the National Coordination of Economic Smallholder Organisations in Bolivia - Coordinadora de Integración de Organisaciones Económicas Campesinas. CIOEC has a membership of seven commodity based federations and 64 primary economic organisations, bringing together an estimated 80,000 producer households.[112]

3.1 Sistema Boliviana de Innovation y Transferencia Tecnológica (SIBTA)

SIBTA was designed in 1998 and replaced the discontinued public national agricultural research system (IBTA) with a modern system geared towards value chain development and the subcontracting of R&D to private entities. It established four Foundations for Agricultural Technological Development (FDTAs) as operational branches in the different Macro Eco Regions: Valleys, Altiplano, Trópico Húmedo and Chaco. These FDTAs are autonomous units that directly manage international donor support without direct interference by the Ministries. FDTA-Valleys was the first foundation to start activities, funded largely by USAID. FDTA-Valleys uses a sophisticated and innovative form of resource prioritization in R&D. It collects the requests (*perfiles de proyectos*) from legally constituted organised actors in the value chain, including producer organisations, and organises a review and assessment of these requests by anonymous external panels. In order to make a selection from the requests submitted, a public procurement call is placed to motivate private R&D institutes to elaborate full proposals. Based on these proposals an external review panel qualifies the project and, when a project has been selected, a formal 'no objection' statement is required

---

[112] The author worked between 1999 and 2004 as an economist in CIOEC-Bolivia, supported by ICCO.

from the requesting organisation. An estimated 126 to 390 days[113] after the initial request, the R&D arrangement can begin.

Farmers' organisations responded positively to this new mechanism for agricultural R&D. Instead of farmers articulating research requests directly to bureaucratic and poorly financed NARIs, or submitting them to NGOs where they are subject to trends in development cooperation, the FDTAs are a more sustainable institutional set-up for agricultural R&D. The Board of Directors of the FDTA-Valleys (60% private, 40% public) included several people from smallholder farmers' organisations, and managed to make decisions by consensus, away from the traditional political interference in funding decisions. Co-financing exigencies were stringent, and were the main concern for farmers in the start-up phase. However the issue became secondary when co-financing arrangements with other support agencies, municipal authorities in particular, were admitted. The focus of producer organisations in discussions on SIBTA shifted from access to R&D towards governance of R&D. Specifically the process of contracting R&D service delivery became contentious.

Farmers' organisations that submitted request proposals felt uncomfortable with contracts signed by FDTA as one party, and the R&D agency as the other. They suggested a different modality: a R&D contract between the producer organisation and the R&D agency, with FDTA as a third party. The idea behind this proposal was the realistic assumption that the R&D intervention would probably be imperfect from the start and that adjustments would have to be made during the course of the program. The POs, as the requesting party, wanted sufficient power to eventually push for changes in the R&D project if needed to generate more useful R&D products. A related point was that of property rights to the assets acquired during the R&D project. Producer organisations claimed the right to be final beneficiaries of investments made in the project as a reward for their financial contribution. FDTA contract conditions made this transfer of assets conditional on a positive evaluation by FDTA, and not an inherent right. Obviously, this restricted the negotiating power of the producer organisations relative to FDTA in the case of conflicting views during the R&D arrangement.

The salience of both points surfaced on one of the first projects to start with FDTA support. The Oregano project submitted to FDTA-Valleys in 2002 was elaborated by the second-tier cooperative AGROCENTRAL, in close cooperation with the Canadian agri-agency SOCODEVI. FDTA-

---

[113] A time lapse that depends on the donor involved: BID or UDAID (Jackson and Wing, 2003)

Valleys co-financed investments and staff for a project to establish oregano production in five cooperatives around the village of Tomina in the Chuquisaca Department, and implement state-of-the-art drying facilities to meet export quality (Paz and Céspedes, 2005). The oregano project had a kick-start, as the project was considered to be of strategic interest, and built on several years of R&D intervention in the chain with SOCODEVI support. As an exception to the rule, the R&D project was directly contracted without a public bidding process. This was motivated by the need for short term results, necessary for FDTA external communication[114]. The Oregano Project started in 2003. AGROCENTRAL itself was the requesting and the supplying partner of the R&D project. During the project several points of contention arose between the FDTA-Valley and the producer organisation. The initial governance structure that placed AGROCENTRAL as chain coordinator, accountable to the five member cooperatives, was gradually replaced by a structure wherein chain coordination and value distribution became concentrated in a company directed by hired staff, directly financed and accountable to FDTA-Valleys. Board members of AGROCENTRAL were gradually bypassed on crucial aspects of decision making by professional staff paid by FDTA.

Directly after approval of the R&D project, FDTA tried to replace AGROCENTRAL in oregano marketing and pressed for the creation of a specialized and autonomous company, UNEC. It was proposed as a more professional, flexible and versatile institutional set-up to coordinate chain activities, do oregano marketing and manage input supply. The marketing capacities of AGROCENTRAL were considered insufficient for short term commercial success. Production was seen as the sole task for the farmers and their organisation. R&D, exporting and marketing were regarded as tasks that necessarily had to be delegated and subcontracted to professional and specialized units that could make decisions independent from the farmers. The initial idea of FDTA was to have the five first-tier cooperatives as minority shareholders of the company, in addition to FDTA and SOCODEVI, and without AGROCENTRAL.

AGROCENTRAL board members felt extremely uncomfortable with these dynamics. They considered it a breach of contract and partnership

---

[114] Funding by USAID to FDTA-Valleys was initially limited to the first two years, with further support contingent on a positive evaluation of FDTA institutional development. This created the institutional necessity to generate positive outcomes and tangible results in a relative short time.

and a threat to the viability of AGROCENTRAL as an institution. However, they were afraid of direct confrontation with FDTA officials, due to a lack of detailed knowledge on alternative forms for institutional arrangements, and for fear of being bypassed as owners of the established processing units after completion of the first phase of the R&D arrangement. Decision making within AGROCENTRAL had come to a point where the project was going to be aborted. The oregano case was discussed and questioned in a regional platform of economic peasant organisations organised by CIOEC and SNV. These discussions motivated a quest for detailed knowledge and capacity building on different modalities for peasant enterprises by AGROCENTRAL and other organisations. In 2004, AGROCENTRAL asked its national (third-tier) organisation CIOEC to develop an alternative proposal for UNEC governance that would respect AGROCENTRAL ownership of the original R&D demand and generate conditions for increasing marketing efficiency. CIOEC had accumulated broad experience with the pros and cons of separating the marketing activities of producer organisations into a specialized branch. After several meetings with the AGROCENTRAL board members, a 'counter-proposal' emerged for a division of UNEC shares between AGROCENTRAL and cooperatives, while inviting supporting donors to be a part of the executive board, but not shareholders of the company. After heated debate the proposal was accepted by SOCODEVI and FDTA, and resulted in the formal establishment of UNEC as a limited liability company with FDTA, SOCODEVI and AGROCENTRAL each having 33% of the votes on the board.

3.2 Servicios de Asistencia Técnica para Pequeños Productores (PROSAT)

PROSAT is a Bolivian R&D pilot project that explicitly refers to producer organisations as its prime target group. The idea behind this IFAD-sponsored project is to boost demand driven support to groups of farmers by creating a market of R&D service providers for smallholders. PROSAT started before SIBTA restructuring, and was embedded in national and departmental government structures. PROSAT is a pilot co-financing mechanism in which the financial support to R&D work from the beneficiaries' own resources is gradually increasing from 10% to 60% over a 2-3 year period. PROSAT worked quite well with established producer organisations that had donor support through development cooperation. Smallholder organisations without such support, however, were generally not able to pay such large amounts. The increase in members' income was generally not enough to cover the additional expense. Even if this individual income increase could in some cases have been channelled towards paying the PROSAT co-financing requirements, this would have had such a negative impact on members' enthusiasm that the project would have surely collapsed. As such, other 'creative' arrangements to keep the R&D arrangement going had to be found. The system lead to the appearance of isolated and loosely articulated groups of beneficiaries that emerged around individual unemployed extension officers (CIOEC, 2000). In these projects the contribution of the beneficiaries was generated by a 'hidden' contribution from the officers' salary. In many cases, the officer him/herself formed the association around personal contacts in a village, neither requiring nor expecting a monetary contribution. Sometimes, the emergence of these new associations affected the support base of existing organisations in the area that was providing similar services. Theoretically, the induced competition between producer organisations could positively influence the quality of services in their constituency. However, this did not happen in practice as these services were generally geographically targeted, totally based on donor support and executed under conditions agreed upon with donors. Quality changes observed were more due to donors learning from experience than to competition for membership. Donor coordination could have been far more effective if the focus had been on improving the quality in service provisioning to the existing producer organisations rather than eroding their member base by setting up alternative organisations.

PROSAT effectively lead to increased beneficiary control and ownership over R&D support. But it had limited impact on chain development, as most projects focused on production increases without proper analysis of market conditions (IFAD, 2005). Many prospective new crops or rural processing activities have been supported with R&D, without developing a sustainable market outlet to form the basis for upgrading pilot experiments and paying even a minimal salary to the R&D officer. CIOEC signalled several of these shortcomings in communication with government officials involved, without positive results. Farmers' organisations were neither represented nor consulted during the project appraisal in PROSAT nor in the monitoring of results. The active involvement of farmers' federations in screening proposals could have been useful to target R&D support to organisations with a minimum of collective marketing experience. The main argument for excluding the farmers' federation from the R&D system was that they would otherwise both assist member organisations in submitting proposals and be involved in evaluating them. This potential conflict of interest could have been easily resolved if the differences between types of farmers' organisations had been properly understood and had roles been divided correspondently: grassroots farmers' organisations for contracting R&D services; and farmers' federations for controlling the R&D provision process.

4. Conclusions

Smallholders in developing countries need to organise themselves in order to become attractive business partners in value chains. Therefore, some smallholders have created farmers' organisations that engage in collective marketing or processing activities. To survive competition with other national or international value chains that often serve the same markets, R&D is needed to improve production and to increase performance of the value chain. R&D service providers increasingly try to capitalize on the relative organisational strength and financial autonomy of farmers' organisations that are active as chain operators. Instead of starting parallel organisational structures for the provision of specific services, as was done in the past, service providers tend to concentrate support services more and more within established economic farmers' organisations.

Competitive funding appears to be a good way to make R&D more responsive to the demands of farmers' organisations. It is flexible and allows incorporation of the R&D needs of farmers' organisations in production but also further downstream in the value chain - in processing and exporting. There are many different ways to involve farmers' organisations in R&D systems. Some systems make it easier for them to perform their roles properly than others. It is important to distinguish different types of farmers' organisations as they may play different roles within the R&D system. Coordinated involvement of these different types of organisations in the national R&D system may prevent frustrations. Often overly high expectations are placed on either grassroots organisations or on national federations. Grassroots organisations and commodity based federations can play a crucial role in articulating R&D demands, while regional and national farmers' federations are especially useful in monitoring the R&D process. These federations can support their members in negotiations with private service providers. They can monitor the R&D process and mediate where there are conflicts of interest between farmers' organisations and private service providers. To facilitate this monitoring role, both the designers of R&D systems and farmers' federations need to build knowledge and experience as to possible and effective ways of arranging R&D. Therefore, it is important to open up effective channels of communication and learning between the designers of R&D systems and farmers' federations.

References

Alex, G., D. Byerlee, M-H Collion and W. Rivera, 2004. Extension and Rural Development: converging views on institutional approaches? ARD Discussion Paper 4. Washington, World Bank.

Alex, G. and W. Rivera (eds.), 2004. Extension Reform for Rural Development: Case Studies of International Initiatives Volumes 1-5. ARD Discussion Paper 12. Washington, World Bank.

Bosc, P., D. Eychenne, K. Hussein, B. Losch, M.-R. Mercoiret, P. Rondot and S. Mackintosh-Walker, 2001. Reaching the Rural Poor: The Role of Rural Producers Organisations (RPOs) in the World Bank Rural Development Strategy - background study, World Bank.

Byerlee, D. and R.G. Echeverria, 2002. Agricultural Research Policy in an Era of Privatization, CABI Publishing.

CIOEC, 2000. Agenda para el Desarrollo Estratégico de las Organisaciones Económicas Campesinas, CIOEC-Bolivia, La Paz

CIOEC, 2003. Cadenas Productivas y Agricultural Campesina: interrogantes al Sistema Boliviana de Tranferencia Tecnológica, La Paz, CIOEC

Echeverría, R.G., E.J. Trigo and D. Byerlee, 1996. Institutional Change and Effective Financing of Agricultural Research in Latin America. World Bank Technical Paper no. 330. Washington, Word Bank.

Hall, A., W. Jansen, E. Pehu and R. Rajalahti, 2006. Enhancing Agricultural Innovation: How to Go Beyond the Strengthening of Research Systems. Washington, World Bank.

Hussein, K., 2001. Farmers' Organisations and Agricultural Technology: institutions that give farmers a voice, ODI.

IFAD, 2005. Bolivia- Country Programme Evaluation, Rome and La Paz, IFAD Office of Evaluation

Jackson, D. and H. Wing. Evaluation of the Market Access and Poverty Alleviation: MAPA Project in Bolivia. Washington D.C., USAID/Bolivia

Muñoz, D., 2004. Small Farmers Economic Organisations and Public Policies; a comparative study, La Paz, Plural Editores

Paz, A. and L. Céspedes, 2005. Exalibur: cadenas productivas para luchar contra la pobreza. Cochabamba, DFID-PROINPA.

Ton, G. and J. Bijman, 2006. The Role of Producer Organisations in the process of developing an integrated sypply chain: experiences from quinoa chain development in Bolivia, In: J. Bijman, O. Omta, J. Trienekens, J. Wijnands and E. Wubben (eds.). International agri-food chains and networks: management and organisation. Wageningen Academic Publishers, the Netherlands, pp.97-111.

Ton, G. and D. Jansen, 2007. Farmers' organisations and contracted R&D services: service provisioning and governance in the coffee chain. Markets, Chains and Sustainable Development Strategy & Policy paper 4, Wageningen.

Wennink, B. and W. Heemskerk, 2005. Farmers' organisations and agricultural innovations: case studies from Benin, Rwanda and Tanzania. Amsterdam, KIT.

Williamson, O.E., 2002. The Lens of Contract: applications to economic development and reform. Forum Series on the Role of Institutions in Promoting Growth. Maryland, IRIS-USAID.

Worldbank, 2004. Agricultural Investment Sourcebook. Washington.

Producer organisations and market chains

Contract farming and social action by producers: the politics and practice of agrarian modernisation in the Philippines

Sietze Vellema

This chapter describes the role of producer organisations in the Philippines, particularly their role in contract farming schemes. Surprisingly, POs have to date played a minor role in contract farming discussions and negotiations. This limited role can be explained by the political background of most POs: they have mainly addressed the rules and practices of land reform.

1. Introduction

Contributions in this volume capture the changing roles and functions of producers' organisations, often based on working directly with organised farmers or peasants. This chapter concentrates on a specific organisational arrangement in agri-food chains, namely contract farming, and examines how such an arrangement, which is an outcome of corporate strategies in global markets, is embedded in the social and political dynamics of agrarian modernisation in the Philippines. The politics and advocacy work of farmers' organisations had an important influence on shaping these dynamics. Firstly, I will describe how contract farming was induced by Philippine agrarian reform policies. Farmers' organisations have been very active in the struggle for land reform, and the question is how they anticipate or react to new agricultural policies and institutional arrangements emerging from the political aspiration for land reform. In Mindanao, the most southern island in the Philippines, the issue of contractual arrangements between small and medium sized landowners on the one hand and agribusiness and food companies on the other, have become important elements in official land reform programmes. Secondly, I will examine the 'real' functioning of contractual arrangements in Southern Mindanao. I illustrate that the lobby work of farmers' organisations for agrarian reform has been strongly biased towards land distribution to individual landowners and that these lobby strategies were not able to capture the power relationships and interdependencies intrinsic to market linkages and contractual arrangements between producers and buyers in an agri-food chain. The analysis identifies a number of entry points where collective action of organised producers could improve the position of farmers in contract farming arrangements. The contribution builds on an in-depth analysis of two contract farming schemes in the Philippines (Vellema, 2002; 2003; 2005).

2. Politics: export agriculture, agrarian reform and contract farming

Equitable (re)distribution of land has been the strongest demand of advocacy and action within farmer and peasant movements in the Philippines. The Spanish colonial history of the Philippines resulted in large sugar cane and coconut plantations. In Mindanao, large scale, export-oriented plantations, mainly pineapple and banana, were established more recently during the American colonial period and after World War II and independence. After the popular uprising in 1986 that toppled the dictatorial regime of President Marcos from power, various agrarian reform advocates placed the contribution of export agriculture to national industrialisation and regional growth, high on the political agenda. Agrarian reform policies adopted by the Philippine government were strongly influenced by the outcomes of negotiations and the strategies of peasant organisations and NGOs rooted in the persistent struggle for land redistribution.

Although the tendency towards export-oriented development of small and medium-sized farms was reflected in government policy since the Marcos era, farm businesses and agribusiness in Southern Mindanao didn't really take off until the late 1980s. At the same time, as a result of the overthrow of the Marcos regime and the return to elected power, government policy took up the challenge to start an ambitious land reform programme. This was combined with parallel efforts in several regional development plans that focused on further integrating Mindanao's agricultural sector into international markets. A strict implementation of the new land reform laws would have had serious consequences for traditional landlords and for multinational corporations leasing public land. Understandably, the reform came up against fierce opposition from the land owning classes and, more in a more restrained manner, from international agribusiness firms.

At the end of the 1990s, former Secretary of Agrarian Reform, Ernesto Garilao, stressed the possible link between land distribution and the improvement of agricultural productivity. In his opinion a lot had already been done on equity and land distribution so that the real battle had become productivity improvements: *'The poor must have some productive assets - land, credit, education, etc. So, if you are able to provide him land, the next is, how would you help him become productive so he can have a higher income. Hopefully, if he's going to be a good farmer, he gets out of poverty.'* (Manila Chronicle, 03/17/96). During the years when Horacio Morales with a strong background in NGO advocacy work on land reform, was Agrarian Reform Secretary under President Estrada, the problem of raising productivity and profitability for farmers led the agenda.

Over the years, policy makers have considered various arrangements between farmers and agribusiness for creating favourable conditions to improve the productivity of small-scale farms. Contract growing, amongst others, has become an institutional arrangement advocated by many policy makers. *'What we do not like to see is another wave of plantations'*, Garilao told Business World (01/09/95). *'We always feel it [contract growing] can be done on small farms, where farmers bring their produce to the plant'*. Garilao was the person who positioned contract growing within a new paradigm, namely creating competitive agriculture. It is no longer feasible to give land just to a tenant or to farm workers, he stated. *'Farmers' associations and cooperatives have to be related to markets, you need links to agribusiness. But, the conditions of the arrangement should be favourable for both sides. The government should assist in realising such a symbiotic relationship*' (interview: 1995). In general terms, advocates argue that contract farming for labour-intensive crops should bring scarcity of land and abundance of labour under a coherent umbrella for regional economic development (Hayami *et al.*, 1990, Adriano, 1988). These ideas have been applied across the Philippines, but contract farming was introduced especially on the island of Mindanao. Mindanao appeared to be well-situated in terms of inducing change within the agrarian structure (APRAAP, 1995).

The actual implementation of the Philippines land reform programme in Mindanao accorded a major role in agrarian modernisation to agribusiness and international food corporations. The persistent specialisation of the region towards agro-export production began during the American colonial era (Boyce, 1993) when the interests of large landowners were firmly established (Tan, 1995; Azurin, 1996). In its modernisation policies, the Philippine government combined enhancing productivity in agriculture with sustaining normal agribusiness operations (de Lange, 1997). Likewise, management of multinational food companies operating in Mindanao, such as Dole and Del Monte, played a key role in the government's endeavour to gain public and political support for its reform programme. Agribusiness and food corporations could comply with land reform legislation by redistributing land ownership to cooperatives formed by their employees. The workers' cooperatives then engaged in a arrangement whereby they leased the land now formally owned by the cooperative to the company, and consequently generated income for the cooperative members. Dole and Del Monte agreed to such a leaseback arrangement. In fact, their pineapple plantations were among the first to be covered by the land reform programme. Since then, Dole has had to negotiate the terms and duration of its lease agreement with the newly formed cooperative. Simultaneously, however, the company continued its expansion by leasing land from private owners. Interestingly, under the new law, there was no limit as to the amount of land that a foreign or domestic agribusiness corporation could lease from local landowners (Putzel, 1992).

Mindanao became a forerunner in the creation of institutional linkages between internationally oriented agribusiness and independent farmers, which received substantial support from international donor organisations, such as the World Bank and the Asian Development Bank.

In the 1970s, the US government was already prioritising large-scale agribusiness ventures (Williams and Karen, 1985). The US government was biased against distributive land reform and focused instead on increasing agricultural productivity by introducing new technologies, abolishing tenancy and exploring new frontiers (Putzel, 1992). In the early 1990s, USAID was the main financier of a major development programme for the region, which built the infrastructure necessary for the expansion of export agriculture (Tiglao 1994; McAndrew, 1993). Central notions in the various development efforts in this province were infrastructure development, strengthening of growth clusters, involvement of the private sector, diversification in export-oriented production, and labour intensive agro-industrialisation (Llorito, 1995a, 1995b, 1995c). With this support, contract growing became a showcase for the introduction of new, high value crops, the creation of marketing facilities, and the coordination of farming practices with the harvesting and processing operations of an agro-industrial corporation (USAID, 1999). This support was motivated both by a market-led view of economic development and by the desire to create peace in a region with a violent history of rebellion and armed conflict. The political history of producers' organisations in Mindanao was largely shaped by this combination of agribusiness expansion and civil conflict and rebellion (Vellema and Vervest, 1996). The next section discusses how producers deal with one particular institutional arrangement originating in this process of agrarian modernisation: contract farming.

3. Practice: land, corporate strategies and negotiations

With the rapid economic expansion of the 1990s, agribusiness corporations were faced with the problem of how to secure access to land. This imposed the task of addressing and solving new organisational challenges facing these corporations: contract farming is a widely used means to link producers to agri-food chains. Commonly, contract growing schemes are responsive to existing patterns of land distribution. A company may decide to introduce new production schemes or expand its production facilities, but access to land is still a basic requirement. Vacant tracts of land are rarely available in most parts of Southeast Asia. Furthermore, in the case of the Philippines, agrarian reform programmes hindered the establishment of large scale plantations. Many areas are occupied by a diverse population of smallholders, absentee landlords, plantations, etc. especially those with fertile soils, irrigation infrastructure, or good access to markets or harbours. Hence, agribusiness companies in Mindanao had to deal with existing landholdings.

In general, contract growing is as a way to gain access to land, without necessarily controlling the productive activities taking place on that land. Contract growing can take many forms, even within a single company. In many instances, particularly for crops such as bananas, contract growing can be seen as a way to complement or partially replace plantation agriculture. In most contract farming arrangements, the grower is made responsible for 'assembling' the agricultural commodity. Yet, the company exercises varying forms of control (and varying degrees of coercion) over the production process. Contract farming is usually meant to introduce new, high-value crops, create marketing facilities, and coordinate farming practices with the harvesting and processing operations of an agro-industrial corporation. Contract growing often entails a combination of a production contract with individual producers, and a financing agreement with associated producers. A production contract requires the producer to deliver a particular crop for a price often specified in advance, and contains the condition that technical assistance and advice, given by company technicians, should be followed. Simultaneously, contract growing becomes a credit relationship, including the provision of agricultural inputs and advance payments for labour. This makes contract farming attractive for producers who lack credit and marketing expertise. The provision of credit is often a central element in new linkages between agribusiness and farmers, which makes contract farming an offer most small and medium farmers simply cannot refuse.

Companies can decide to introduce new production schemes or expand production facilities, but in the Philippines, company access to arable land and farm labour was restricted by government regulations. They are therefore forced to enter into complex social relationships with landowners and farm workers. Contract growing arrangements entail a social relationship between the company and the actual owner of the land. A formal understanding of contract farming wrongly assumes that the contract transforms socially and economically unequal parties - the transnational corporation and the small farmer - into political and legal equals. The mixture of actors included in contract growing makes it difficult to decide which parties are in effective control of the scheme and how major decisions are made. A strong feature of contract growing is the capacity to incorporate and coordinate diverse interests and organisational styles into one single organisational framework. In Mindanao, contract farming entailed the integration of a diverse group of landowners into agribusiness systems, ranging from absentee landlords also engaged in other business ventures or local politics, former tenants and settlers who benefited from land distribution programmes, small farmers involved in dependency relationships with wealthier neighbours, or leading Muslim families owning large tracks of land and acting is brokers on behalf of their political constituency. All these farmers had relationships with various actors: input suppliers, traders, banks, processors, landowners and government agencies. Obviously, they have different stakes in a contract growing scheme. The different social positions of contracted farmers affect how they negotiate the terms of the contract and how they relate to organisational and managerial strategies under the arrangement.

The profitability of a contract growing arrangement is closely related to the flow of capital to and from other parts of the agri-food chain. National or international headquarters place very strict guidelines on profit margins. Typically, local management, although confronted by the demands of growers, is not the decision-making party when it comes to price policy due to the hierarchical structure of most agribusiness enterprises. However, some room for negotiation can always be found. The negotiating power of growers is complex; they are aware of that the company is dependent on them for growing the profitable crop but at the same time they are aware that their indebtedness due to credit provisions inherent in contract growing gives them a subordinate position vis-à-vis the company. The design of financial systems and accounting procedures is pivotal in terms of securing a win-win situation: contract conditions related to credit flow, prices, and qualification must be transparent and the terms of the arrangement must be negotiated between the parties involved rather than imposed. Much of the critique of contract growing arrangements centres on this financial dependence and the danger of indebtedness (for example, IBON, 1998). My own research and a number of other case studies (e.g. Collins, 1993, Korovkin, 1992, Glover, 1984, Goldsmith, 1985) confirm that the often asked question whether contract farming is good or bad invites a firm and strong opinion on the contractual arrangement *per se*, which disregards the actual practice of contract farming (Little, 1994). Contract growing allows reward for effort and performance; it creates room for small and medium farmers to enhance their farming business and to invest in, for example, education for their children. Moreover, growers recognise and use opportunities

Producer organisations and market chains

in the management system to negotiate their interests by building on the social organisations they have been engaged in, such as family networks, strong local communities with brokers, political or business networks, labour unions, or reciprocal relationships with technicians. The gradual acceptance and expansion of contract farming can be a result of farmers' capacity to respond to favourable development policies and market opportunities (Little and Watts, 1994). Little (1994) emphasises the importance of assessing contract schemes in the light of realistic alternative options for increased production and income (cf. Porter and Phillips-Howard, 1995). Pre-existing checks and balances, whereby farmers have recourse to judicial procedures, laws regulating contractual issues (e.g. risk), or an institutionalised bargaining position inside the company structure, are often missing.

The relationship between a company and its growers crystallises around remuneration, i.e. price. Growers decide what income is acceptable and when price negotiations with the management are necessary. A grower might decide to act individually and secure benefits. Growers can also decide to enter into collective bargaining. Different kinds of growers may have different appraisals influencing how they act, and, hence, have different opinions and strategies towards the company. My research showed that company employees tried to limit the negotiating power of producers' associations organised under the contract farming scheme. I observed several organisational and political efforts to obstruct collective bargaining as well as a shift towards personalised relationships between growers and company technicians. Moreover, the existing form of collectivity in the scheme, an association, was a condition for obtaining credit from the bank and not embedded in any social organisational structure.

In principle, producers' organisations can play a constructive role in the financial management of contract growing schemes, possibly in alliance with semi-public financial organisations that are frequently providing support. Surprisingly, however, the government institutions co-financing the modernisation programmes in Mindanao did not have an explicit policy to avoid lopsided relationships between growers and the company Advocates of contract farming ceaselessly emphasise the potential of this institutional innovation in order to overcome imperfect market conditions. And indeed, grower incomes often do rise. Growers have less trouble marketing their produce, and company technical assistance makes a substantial effort to improve farming practices. However, my research shows that contractual arrangements are less transparent in cases of conflict or business troubles. In those situations, the specific terms of the contract come to the fore. Production costs may rise, prices and quality standards alter, or the company's financial policy becomes less accommodating. Also the semi-public financial organisations involved in the schemes examined did not pay much attention to the specific terms of handling failure and managing risks. Collective loans and credit lines were offered to loosely associated growers, and when financial problems became apparent, financing organisations tried to make these associations collectively responsible for the failure of individual members to amortise their loans. Philippine cooperative law to some extent sets this up but the legal implications for associations were less defined. In such a situation, a producers' organisation can play a strong role in setting the terms, recognising that marrying farmers to firms may also end in divorce.

To further understand the position of contract growers in an agri-food chain, we have to gain insights as to the 'institutional hierarchy' of the specific company. Coordination of various tasks, as well as control and coercion are factors shaped by both the actions and needs of the company and the bargaining power or compliance of growers. It is relevant to distinguish those elements that can be negotiated by growers, and those which are beyond their control. We have seen that this depends on the financial capabilities of the growers. A grower who is able to finance production by him/herself will likely have more bargaining power than a grower who relies on credit provided by the company. The financial capacity of a grower and the diversification of his economic activities, as well as the way they in which he organises labour and manages the farm, are factors that shape the interest growers have in collective negotiations. Nevertheless, when growers experience problems due to the behaviour or interventions of the company, political differences may suddenly cease to exist. The contract itself is insufficient to unify a diverse group of growers; collective action is especially induced by struggles around specific practical issues and day-to-day interactions with the company. My research showed that growers still have substantial influence in managing tasks affecting performance in terms of quantity and quality of production, which makes them a valuable element in the corporate strategy (Vellema, 2002). Nevertheless, there is clear evidence that small farmers who maintain alternative opportunities for production and income are in a much stronger position than farmers whose land area is devoted entirely to the contract crop (Porter and Phillips-Howard, 1997).

Producer organisations and market chains

Producer organisations and market chains

This in-depth analysis of integration through contract farming revealed a diversity of negotiation styles adopted by producers. This produced a mixture of management styles and social preferences in the single organisational form. Some growers, with experience in plantation or factory work, opted for negotiations styles grounded in the collective bargaining processes present in a hierarchical management structure. Other growers considered themselves to be independent entrepreneurs closing a business deal with the company and they negotiated for freedom to manage their farms according to their own insights and interests and, in the case of conflict, considered leaving the contract farming scheme, turning instead to other business opportunities. Still other growers relied heavily on the negotiation capacity of a broker from their community, who often played an influential political role in local constituencies also. Lastly, some growers chose to enter into reciprocal relationships with company employees in order to protect their own interests or to receive particular benefits. Thus, in one single contract growing scheme, producers may have varied social positions and adopt distinct negotiation styles. The consequences of these diverse social positions are reflected in the actions and effectiveness of cooperatives and associations in contract farming. Although growers seem to be in a similar position - each have signed a contract with the same company - their political positioning and behaviour in negotiations reveals differences. One can imagine that a grower who has other business interests in a nearby city and has been elected as city councillor for two consecutive terms, is in a different position than a small scale farmer who planted all of his land with the new crop and relies entirely on contract growing for his income. Furthermore, long-standing political cleavages will also be reflected in collective bodies, although I observed that even deeply rooted conflicts, for example between Christian and Muslim communities, could be overcome through strong bargaining efforts with the company.

4. Conclusion

In this chapter, contract farming is presented as an outcome of the combination of agrarian reform policies and corporate strategies to make land access possible. Contractual arrangements have been introduced with the idea that linking landowners to markets and agribusiness improves the opportunities for agrarian producers to earn decent incomes. In the Philippines, farmers' organisations and NGOs have been very active in shaping agrarian reform policies, primarily demanding land distribution. Remarkably, producers' organisations seemed to be quite absent in these contract arrangements. In plantations affected by agrarian reform, trade unions played a pivotal role in defending collective interests and in some cases, even became shareholders leasing land to large corporations. In the contract farming schemes that were integrating small and medium landowners, the level of collective action in defending economic interests was low.

This is remarkable because a win-win situation is not an automatic outcome of contractual arrangements, as is suggested by the advocates discussed above. Whether a contractual arrangement can yield material benefits for contractors depends largely on the specific terms of the arrangement, the forms of governance and administration, and on the working relationships operating in the scheme rather than the fact of contracting *per se*. Thus, it seems to be appropriate to examine what happens inside the organisation and to ask whether contract farming can be managed differently. In many development efforts, attention focuses primarily on making the actual connection and less effort is put into managing the relationship. Making producers' organisations part and parcel of managing schemes to integrate independent farmers into agribusiness systems appears to be an uncharted terrain.

The chapter illustrates that the behaviours and strategies of producers are socially embedded. The contract growing schemes examined included a variety of forms related to the social and historical backgrounds of the farmers involved. The company tried to cater to this institutional variety. Simultaneously, this variety hampered the establishment of a producers' organisation defending the collective interests of contracted producers. Social organisations in Mindanao were rooted in specific social and political struggles, such as land reform, labour rights, and movements striving for the regional autonomy of Muslim Mindanao. These social organisations seem to be less equipped to take part in complex negotiations and interactions in contract growing schemes.

The suggestion of this chapter is to take political history and varied social realities as a starting point for reconfiguring market linkages and chain integration. This may lead to a variety of institutional forms, grounded in specific social conditions, providing producers with effective tools for managing the conditions of modernization differently. Such an approach may contrast with a strong focus on organisational fixes, such as contract farming or cooperatives, which can be observed in the development cooperation and market-led development debate. The chapter proposes a search for possible linkages between broader social struggles and enabling policies in the domain of agrarian reform, with functional organisational arrangements in an agribusiness development scheme.

References

Adriano, L.S., 1988. Agribusiness and Small Farmers: Partners in Development. Manila Chronicle, August 30, 20.

APRAAP, 1995. Mindanao 2000 development framework plan. Agricultural Policy Research and Advocacy Assistance Programme, Office of the President for Mindanao, DMJM International Inc.

Azurin, A.M., 1996. Beyond the cult of dissidence in southern Philippines and wartorn zones in the global village. Quezon City: Center for Integrative and Development Studies, University of the Philippines Press.

Beckett, J., 1994. Political families and family politics among the Muslim Maguindanao of Cotabato. In: A.W. McCoy (ed.) An anarchy of families: state and family in the Philippines. Madison: The Center for Southeast Asian Studies University of Wisconsin, p.285-309.

Boyce J.K., 1993. The political economy of growth and impoverishment in the Marcos era. Quezon City: Ateneo de Manila University Press.

Collins, J.L., 1993. Gender, Contracts and Wage Work: Agricultural Restructuring in Brazil's Sao Francisco Valley. Development and Change, 24, 53-82.

Glover, D.J., 1984. Contract Farming and Smallholder Outgrower Schemes in Less-Developed Countries. World Development, 12 (11/12), 1143-57.

Goldsmith, A., 1985. The Private Sector and Rural Development: Can Agribusiness Help the Small Farmer? World Development, 13 (10/11), 1125-38.

Hayami, Y., M.A. Quisumbing and L.S. Adriano, 1990. Toward and Alternative Land Reform Paradigm: A Philippine Perspective. Quezon City: Ateneo University Press.

IBON, 1998. Contract growing: intensifying TNC control in Philippine agriculture. Manila: IBON Databank and Research Center.

Korovkin, T., 1992. Peasants, Grapes and Corporations: The Growth of Contract Farming in a Chilean Community. Journal of Peasant Studies, 19 (2), 228-54.

Lange, W.A. de, 1997. CARP coverage of commercial farms. Business World, 14, 15 & 16/08/97.

Little, P.D., 1994. Contract farming and the development question. In: P.D. Little and M. Watts (eds). Living under contract: Contract farming and agrarian transformation in Sub-Saharan Africa. Madison: University of Wisconsin Press: 216-47.

Little, P.D. and M. Watts, 1994. Introduction. In: P.D. Little and M. Watts (eds), Living under contract: Contract farming and agrarian transformation in Sub-Saharan Africa. Madison: University of Wisconsin Press: 3-18.

Llorito, D.L., 1995a. Mission: build Mindanao Incorporated now!' Mindanao Update, 1 (1), 1-7.

Llorito, D.L., 1995b. What makes Mindanao thick?' Mindanao Update, 1 (3), 1-4.

Llorito, D.L., 1995c. A different kind of war for Mindanao'. *Mindanao Update*, 1 (6), 1-5.

McAndrew, J.P., 1993. Aiding inequality: the General Santos City Project in the Philippines. Washington D.C.: Philippine Development Forum.

Porter, G. and K. Phillips-Howard, 1995. Farmers labourers and the company: exploring relationships in a Transkei contract farming scheme. Journal of development studies, 32 (1), 55-73.

Porter, G. and K. Phillips-Howard, 1997. Comparing contracts: an evaluation of contract farming schemes in Africa. World Development, 25 (2), 227-38.

Putzel, J., 1992. Captive Land: The Politics of Agrarian Reform in the Philippines. New York: Monthly Review Press/ Manila: Ateneo de Manila University Press.

Tan, S.K., 1995. The socioeconomic dimension of Moro secessionism. Quezon City: Center for Integrative and Development Studies, University of the Philippines (Mindanao Studies Report no.1).

Tiglao, R., 1994. Growth Zone. Far Eastern Economic Review, February 10, 40-45.

USAID, 1999. Congressional presentation 1999. (www.info.usaid.gov/pubs/cp99/ane/ph.htm).

Vellema, S., 2005. Regional Cultures and Global Sourcing of Fresh Asparagus. In: Fold, N. and Pritchard, B. (eds) *Cross-Continental Food Chains:* Structures, Actors and Dynamics in the Global Food System, London, Routledge: 124-136.

Vellema, S., 2003. Management and Performance in Contract Farming: the Case of Quality Asparagus From the Philippines. In: S. Vellema and D. Boselie (eds) Cooperation and Competence in Global Food Chains: Perspectives on Food Quality and Safety. Maastricht, Shaker:157-190.

Vellema, S., 2002. Making Contract Farming Work? Society and Technology in Philippine Transnational Agribusiness. Maastricht, Shaker Publishing (Wageningen University PhD-thesis)

Vellema, S.R. and P. Vervest, 1996. Echo's uit het verleden: de geschiedenis van de progressieve beweging in Mindanao. Derde Wereld, 14(4), 252-269.

White, B., 1997. Agroindustry and contract farmers in upland West Java. Journal of Peasant Studies, 24 (3), 100-36.

Williams, S. and R. Karen, 1985. Agribusiness and the Small-Scale Farmer: A Dynamic Partnership for Development. Boulder/London: Westview Press.

Conclusions

Jos Bijman, Giel Ton and Joost Oorthuizen

Growing attention for POs ...

The issue of rural producer organisations has been gaining attention from policy-makers, academics and consultants interested in supporting the social and economic position of (smallholder) farmers in developing countries. Donors and NGOs in development cooperation have (re)discovered the importance of POs for rural development in general and for strengthening smallholder access to markets, in particular. The renewed attention to agriculture in the quest for poverty reduction has also lead to a focus on organisational structures that can help improve both the production and marketing of agricultural products. POs are believed to be able to solve some of market access problems, such as lack of market information, thus providing farmers with (more) market opportunities, and thereby strengthening their livelihood.

... for instance in World Development Report 2008

The World Development Report 2008 (WDR2008), entitled 'Agriculture for Development' is a clear example of this renewed focus on agriculture and the role of POs in supporting agricultural development. The report, incorporating earlier World Bank studies on POs (e.g., Rondot and Collion, 2001) argues that POs are a major part of institutional reconstruction, using collective action to strengthen smallholder positions in (factor and product) markets. POs can reduce transaction costs, strengthen bargaining power, and raise the voice of smallholders in the policy process.

POs are very diverse ...

While POs share a number of characteristics, such as being member-based organizations, the reality is that there are still many different types of POs. A number of PO typologies have been developed based on the formality/informality of the organization, functions, geographical scope, or legal status. The WDR2008, for instance, distinguishes between POs on the basis of three clusters of functions: advocacy, providing economic services, and more general support to local development. Very often a PO combines different functions, although single-purpose POs do exist, such as commodity-specific organisations. In addition, POs often have their main activities at different geographical and political levels.

Both commodity-specific organisations and advocacy organisations often have both local and regional (or national) branches. The common structure for these multi-layer POs is that of a federation (or union), with the lower level organisations being the members of the higher level organisation.

... and so is the institutional environment.

Not only do we find differences between POs, based on function and organisational structure, but institutional environments may also differ substantially. Social traditions, legal structures, policy making processes and market institutions all vary from country to country. As POs are organisations meant to establish favourable linkages between members (rural producers) and outside partners - whether these are private companies or government agencies - POs must adapt to the social, institutional and economic context in which they operate. Changes in the institutional environment, such as shifts in governmental policies or market structure, will have an impact on the efficacy of the PO and may call for revisions in terms of the form and function of the organisation.

Supporting POs requires acknowledgement of their functions ...

With so many different types of POs, no single tool or approach exists to supporting all activities and kinds of organisation. From a capacity building and organisational strengthening perspective, it is important to distinguish between the different functions, as well as between the levels of operation (whether the PO is a local, regional or national organisation). Advocacy groups (as well as commodity groups) at the national level are in particular dealing with national political and governmental issues and structures. Supporting these kinds of POs requires an emphasis on assisting them in developing or refining policy recommendations as well as in developing advocacy strategies. Second tier or regional organisations are more focussed on providing economic and technical services to their grassroots member organisations or they are responsible for processing and marketing members' products. At the local level, the emphasis is usually more on service provision to individual farmers. Technical capabilities as well as sound business practices (including transparency in the costs and benefits of operations) are the most important targets for support to grassroots level POs.

... as well as their basic characteristics

Several lists covering the essential features of POs have been developed over the years. The International Cooperative Alliance (ICA) list of seven principles on which cooperatives are (or should be) based is well known: (1) voluntary and open membership; (2) democratic member control; (3) member economic participation; (4) autonomy and independence; (5) education, training and information; (6) cooperation among cooperatives; and (7) concern for community (see: www.coop. org). Rondot and Collion (2001) state that all POs are characterised by two principles: utility and identity. The principle of utility means POs are useful for members and that members are actively committed to achieve jointly agreed objectives. The principle of identity emphasizes the shared identity within the PO: members usually share a history and a geographical space; members have agreed upon a set of rules that govern internal relations (among members) and external relations (between the PO and the outside world); and members have a common vision of the future, both for themselves and for the group as a whole. This shared identity, which on a higher level may be part of a shared ideology, is a strong social mechanism supporting continued and low cost interaction between the members of the organisation. On a higher societal level, the principle of identity may be fuelled by and in turn contribute to a common ideology of collective producer action.

Most POs are community-based organisations ...

The principle of identity is strongly related to the social and geographical boundaries of membership. Most POs are community-based organisations, operating under community norms and values of social inclusion and solidarity. Members (and often also employees) are drawn from the community and the benefits of the PO directly and indirectly support the whole community. Being a community-based organisation has the advantage of the availability of social capital, which keeps transaction costs low, and a strong sense of ownership, which keeps commitment high.

... but changing market conditions push them to loosen these ties.

Community-based organizations also have disadvantages, such as the multiplicity of the goals, the limited pool of expertise and leadership available in the community, and the entangling of governance of the economic organisation with wider political and social structures such as local hierarchies. Identities within a PO are often not purely based on economic positions or even market orientation. Cultural and political factors can influence the governance of the PO, hampering successful articulation in markets. In addition, well-intended support from NGO's to increase market-oriented capabilities may conflict with the local political or social configuration. Thus the challenge for the PO is to continue to service the community while at the same time becoming more professional in marketing. A possible solution could be for a group of community-based POs to delegate the more commercial and marketing activities to a specialized entity, such as a jointly owned enterprise at the regional level. This federative (or union) PO can than enter into supply chain contracts on behalf of the local POs, while the local POs can continue to combine economic and social functions.

While enhancing market access is becoming more important...

One of the major functions of a PO is to support its members in enhancing market access, either acting as a facilitator, e.g. by providing market information and technical assistance, or as a contractor, by processing and marketing members' products. With the recent trends in globalization and the rise of integrated supply chains, the role of the PO in obtaining market access for members is becoming more demanding. The traditional push strategy - just find a market for every product that the members produce - is no longer feasible. More and more, POs must now align supply and demand by orchestrating member supplies to meet the increasingly stringent requirements of national and international value chain customers. This implies achieving scale and timing in delivery, satisfying sanitary and phytosanitary standards, and meeting the quality specifications of particular processing and retail companies.

... it has implications for management and organisation ...

Strengthening members' market access implies that the PO play an intermediary role between producers and customers (or even the customers' customers). This new or enhanced role of chain intermediary has several implications for PO management and organisation. First, most POs may need to hire external marketing expertise. This will lead to a professionalizing of the management. Second, hiring professional managers in turn requires a strengthening of the governance relationship between management and the board of directors. The members of the board will only be able to direct and control the management when they themselves have sufficient knowledge of marketing strategies and customer requirements. Third, in order to strengthen vertical coordination in the value chain the PO may need to become stricter regarding member compliance with agreements and obligations. POs entering into contracts with (foreign) customers have contractual and moral obligations to deliver the agreed quality and quantity. If members do not comply with quality and other obligations, the reputation of the PO vis à vis its customers and thereby its competitiveness are at stake. Thus, the relationship between member and PO may become more contractual.

... and for the services of the PO ...

Producers that want to supply high-demanding customers, such as domestic and foreign supermarkets, need to improve the quality of their produce, to reduce product quality variability, and to comply with particular quality standards. POs have several tools available to help their members to enhance product quality and thus competitiveness. First, POs may support members in strengthening their competitiveness by providing information on the quality requirements of customers. Part of this function is to assess and choose from the many options for international certification schemes. Second, POs may carry out the sorting and grading of their members' products into quality categories, thus providing an opportunity for incentives to improve quality. Third, POs may provide the technical assistance that members need so as to improve on-farm production methods. Fourth, the PO itself may implement a quality control system, in order to be able to supply customers with a consistent quality. Such a quality control system may enhance producer competitiveness, but also calls for the PO to be strict related to member compliance with the quality requirements. Finally, POs may organize and facilitate innovation processes targeted at reaching higher product quality.

... as well as for ties among the membership.

POs are voluntary membership organisations. Producers only become members when they have a common interest with the other members. Thus, the PO is characterised by homogeneous member interest in the particular function(s) that is performed. However, when the function of the PO becomes more specialized on strengthening market access and facilitating vertical coordination, the interest of members in these new activities may divert. Particularly the interests between small and large (or traditional and modern) farmers become more diverse. Many POs include both small-scale and large-scale farmers in their

Producer organisations and market chains

membership base. Often large farmers are indispensable because they are major users of the PO and thus create the volume in services that allows the PO to be economically viable. In addition, large farmers have the capacity and capabilities to play leadership roles within the PO. When the interests of small and large farmers start to diverge, for instance because large farmers see business opportunities that are not available for small farmers, this heterogeneity can lead to cumbersome and inefficient decision-making processes. A challenge for PO leadership is to find and defend common ground.

Member ownership and member control are crucial characteristics of a PO...

POs are member-owned and member-controlled. Ideally, member-ownership is defined both in economic terms (members are share-holders) and in psychological terms (members feel it is their organisation). Member-control is defined by members holding the decision-rights on both the activities and the investments of the PO. Both ownership and control are collective in nature, i.e., members collectively own the PO and members collectively take decisions as to strategies for the PO. While in theory it is clear why and how these characteristics must be applied, the reality is a bit more complex.

...however, outside support is often needed ...

POs in developing and transition economies often receive substantial support from external stakeholders, such as governmental agencies, donors and NGOs. This support in general is greatly appreciated and in some cases even indispensable to the establishment of the PO. However, POs are and should remain autonomous member-based organizations. This implies that external stakeholders supporting the PO should not take control. Financial and expert support is very welcome but should not become so dominant that the PO is dependent for its very existence and functioning. Even when receiving outside support, strategies and policies decisions should be made by the membership itself. The history of cooperatives in many state-dominated countries has shown that domination by external stakeholders leads to serious problems. It translates into a weak sense of ownership amongst members, which in turn leads to low member commitment. In addition, its leads to low accountability by the board and management of the PO. One of the main challenges for POs that are being supported by governmental agencies or donor NGOs is to remain truly member-based and member-controlled organisations.

... *and should be targeted at empowering producers and their POs.*

Development cooperation agencies, such as donors and NGOs, increasingly recognize the importance of POs in strengthening the position of smallholder producers in national and international agrifood chains. Current trends in these value chains are not favourable for smallholder farmers due to the high and uniform quality and quantity demands of large processing and retail customers. Improving smallholder market access can contribute to poverty reduction, and POs can play a major role in obtaining this goal. Without falling into the trap of POs becoming dependent on outside agencies, support for the empowerment of producers and their POs seems necessary. Such support can consist of several clusters of activities. First, support can focus on the PO itself, such as: capacity building for leaders, members or managers; (other types of) organizational strengthening; building skills to develop and lobby for favourable legislation; building negotiation skills to enter into and maintain partnerships (both vertical and horizontal). This type of capacity building generally is a slow and uneven process that requires donors to be patient and to develop long-term support programmes. Second, development cooperation agencies may help POs to set up market information systems in order to collect, assess and distribute the market information that producers need to improve their competitiveness. Third, outside agents may help POs to provide technical assistance to their members, particularly knowledge and information required to comply with certification requirements.

The role of the government includes general policies to make markets work...

In discussing the role of the government in supporting POs we should make a distinction between direct and indirect support. Indirect support entails strengthening the position of POs in markets. Direct support means, for instance, providing funds for the establishment of a PO or for training PO leadership. Let us focus on the position of the PO in the market and what governments can do to strengthen this position. As markets have liberalized and many state support programs have been abolished, the question arises as to whether there still is a role for governments to play in supporting POs. We think that governments still have an important role, even in liberalized markets. Liberalization of markets does not automatically lead to well-functioning markets. Markets must be regulated and even organized in order to operate effectively. Poorly functioning markets can be characterised by substantial externalities, high transaction costs and unfair trading (i.e., power instead of supply and demand determine the exchange price). Thus, the first task of governments in terms of PO support is to provide a well-functioning market, for instance by providing POs (and all other market players) with information on demand, supply and prices. A public market information system may greatly enhance the efficiency of the market. The second point, and directly related to the first, is the need for law enforcement, which implies a well-functioning judicial system. If the judicial system does not work properly, power determines market outcomes and less powerful market parties may be reluctant to enter into exchanges. Agricultural producers in general belong to the 'less powerful' category.

... as well as specific policies to make POs work.

The third method to support the (working of the) PO is to enact specific legislation. This legislation will provide the PO the legal status necessary in order to enter into contracts and to borrow money: the legal position should be clear, particularly if the PO enters into contracts with supply chain partners. In addition, legislation will determine the basic organizational features of group action, so that the rights and obligations of each stakeholder are clear to everyone involved (particularly PO members). Moreover, it will grant the PO legitimacy as a bona fide organization, which is also a symbol that society acknowledges the importance of this type of economic organization. Finally, special legislation clarifies the basic principles of organized producer collective action, and provides limited liability to members. We like to emphasise here that legislation on POs should be enabling, not restricting. Unfortunately there are many examples of inefficient cooperatives due to restraining legislation.

Favourable policies result from effective lobbying...

Few governments or state agencies are naturally inclined to support rural producers and their POs. Often, policymakers and administrators are more likely to listen to the demands of urban constituencies or large business sectors. Farmers, based outside of the capital city and often not well-organized, are not an influential interest group in policy development and decision-making. The advocacy role of the

Producer organisations and market chains

PO continues to be important, but the content may change due to internationalization of markets and the decreasing sovereignty of nation states to regulate domestic markets. With more emphasis on market access and the (technical) requirements for participation in international supply chains, PO leaders need to be informed on the technicalities of international supply chains and quality assurance schemes. Governments and donors can enhance the effectiveness of PO participation in regional, national and international consultative policy processes by helping them gain access to information and providing funds to recruit expertise to prepare inputs into the policy dialogue and seek professional advice.

... but there are no 'one size fits all' recommendations.

Given the diversity of institutional environment, both in formal (e.g. laws) and informal (e.g. norms and customs) institutions, and the diversity of the economic activities that producers engage in, there is no universal solution to the problem of how to strengthen smallholder market access. Recommendations for empowering producers and their POs should be country-specific and sometimes even group-specific. The chapters in this book have shown that producers and their POs struggle with many institutional, organisational and functional challenges. But they have also shown that there are many potential pathways for improvement. While some of the challenges of group action are omnipresent, for instance the dilemma between economies of scale and commitment, many problems related to market access have arisen more recently as markets have liberalized and (foreign) customers have become more demanding as to quality and delivery conditions. Thus, supporting POs requires recognizing general developments in policies and markets, acknowledging specific organisational features of these types of organisations, and targeting the unique functional characteristics of each local, regional or national PO.

Finally, donors and NGOs should strengthen their collaboration

Within the donor and NGO community there is a lot of experience with POs. Although we have stressed above that there is a large variety of POs and that institutional environments differ, there are commonalities in the management and organisation of POs as well as in the challenges they face. Acknowledging these commonalities opens the window for learning from the experiences of others, and for joint development of pathways for PO facilitation. Particularly the rise of national and international supply chains and the need to help smallholder farmers and their POs to strengthen their position in these chains provides multiple opportunities for collaboration, collective learning trajectories, and joint support projects.

References

Rondot, P., and M.-H. Collion, eds. 2001. Agricultural Producer Organizations: Their Contribution to Rural Capacity Building and Poverty Reduction. Report of a Workshop, June 28-30, 1999, Washington, D.C. World Bank, Washington D.C.

About the authors

Mr. Jos Bijman (PhD) is assistant professor in the Department of Business Administration at Wageningen University in the Netherlands. Before this post he worked as senior researcher at the Dutch Agricultural Economics Research Institute (LEI). His research focuses on economic organization and governance issues in (international) agrifood supply chains, with special emphasis on the role of producer organisations in agrifood chains, the organizational  restructuring of cooperatives, and corporate governance in agricultural cooperatives. He has published on these issues in the American Journal of Agricultural Economics, European Review of Agricultural Economics, and the Journal on Chain and Network Science. In 2006 he edited a book entitled: *International agri-food supply chains and networks; management and organization* (see: www.wageningenacademic.com).

Mr. Kees Blokland (PhD) is managing director of Agriterra, and has an academic background in development economics and cultural anthropology. His PhD field work was undertaken over an eight year period in Nicaragua (1982-90) where he worked through the FAO with the National Union of Farmers and Ranchers (UNAG). From 1990 until 1997 he worked with the Paulo Freire Stichting (PFS). In 1995, his lobby for a rural co-financing  agency was heard by the Dutch Agricultural Board, which led to the founding of Agriterra, in 1997. Five years later AgriCord - the alliance of agri-agencies - became a reality.

Mr. Dave Boselie has many years of experience with international supply chain development for fresh fruits and vegetables in Asia, Latin America and Africa. He has been involved in preferred supplier programs on behalf of leading supermarkets and is currently managing director of AgroFair Assistance & Development Foundation (AFAD). In his current job he aims to promote access to markets for (small) fruit producers through the development of fair-trade and organic business models.

Mr. Dick Commandeur (MSc) has worked for 18 years in Latin America, mainly in the field of rural and economic development. He was an advisor to different producer organizations in Bolivia. After that he led a project for rural development in Nicaragua and over a six year period he coordinated the SNV program for the Andean Valleys region of Bolivia. Currently hc lectures at the University of Sucre, Bolivia on alliances amongst small rural producers.

After studying Business Administration, **Myrtille Danse** (MSc) worked for 10 years in Central America as a business consultant in the field of environmental management, sustainable trade, and integrated chain management (e.g. cleaner production methodologies, international standards such as ISO 9001, ISO 14001, EUREP-GAP and environmental management accounting). In addition, she has worked as a public officer for

the Dutch Embassy on climate change and sustainable trade. Currently she is working for the Agricultural Economics Research Institute in the Netherlands as a Senior Researcher on Sustainable Chain Development and Market Access. An important issue of interest in her work is the stimulation of sustainable inclusion of small producers and small and medium enterprises in developing countries in international supply chains.

Mr. Rik Delnoye (MSc) graduated in Tropical Forestry and Consumer Studies from Wageningen Agricultural University in 1990. He has worked for the FAO, United Nation's Development Program (UNDP) and SNV mainly in Asia and Latin America. He was director of a small enterprise involved in developing and marketing a new range of locally and environmentally produced and processed agricultural products under their own brand name  in the Netherlands. From 2002 onwards he has worked at Agriterra as a liaison officer, responsible for guiding and strengthening partnership trajectories in Asia and Eastern Europe.

Mr. Bo van Elzakker (BSc Tropical Agriculture), Director of Agro Eco, is a consultant for *Agriculture, Crop Protection and Project Management* in the Tropics. Bo has worked internationally - in the Mediterranean region, in the Americas, Asia and Africa - on virtually all aspects of organic agriculture. Since 1994 he has been concentrating on Africa. He is director of the EPOPA program (Export Promotion of Organic Products from Africa,  www.epopa.info), which implements approximately 30 projects with African smallholders to access organic and fair trade markets in the North. Bo can be contacted by e-mail at b.vanelzakker@agroeco.nl.

Ms. Lithzy Flores has worked as an advisor in SNV since 1999. In the programme to strengthen producer organisations in the Andean Valleys, she worked as a specialist in institutional development and organizational strengthening from a gender perspective. She is currently working on entrepreneurship and public-private partnerships for economic development.

Mr. Christian Gouët (MSc) is an agronomy engineer with a BSc in Agriculture Sciences and an MSc in Management of Agricultural Knowledge Systems. He is a guest researcher and PhD candidate at the *Communication and Innovation Studies* Group of Wageningen University. A national from Chile, he gained experience in agriculture, rural development and farmers organisations through work as an extension officer, private consultant  (topics on soil fertility, irrigation and agri-business) and as chief executive officer of a smallholders' associative firm (a commercial and services company owned by farmers unions). In recent years he has undertaken research on the topic of farmers' organisations and development in the context of economic globalization, including international seminars and local case studies in Kenya, Philippines and Chile. Christian is also an international consultant on issues such as the formulation of cooperation programs, farmer organisation financial management, evaluation of capacity building programs, and water governance.

Mr Gerrit Holtland (MSc) is a rural development expert working for the Management Development Foundation MDF (Netherlands), with 20 years of experience. He has worked as an (action) researcher, extension expert and team leader on long term assignments in Uganda, Ivory Coast, Tanzania and Albania. Since 2000 he has been involved as a coach for long term projects in Moldova, Kyrgyzstan, Rwanda and Afghanistan.  Presently his main areas of work are support to producer organisations, Value Chain Development, agricultural strategy design, Public Private Partnerships and organisational and institutional development. Although always employed as a practitioner, Mr Holtland has published widely on (rural) development issues.

Mr. Joost Nelen (BSc) works for the Netherlands Development Organisation SNV in Bamako, Mali. jnelen@snvworld.org. He is an advisor and the key person for SNV's 'Drylands' program. For 13 years he worked as an agronomist in West Africa with farmers' organisations and local authorities on 'rural development' issues such as fair trade, agrarian reform and natural resource management.

Mr. Joost Oorthuizen (PhD) worked as a process manager for the Agri-ProFocus partnership - the Dutch public-private partnership for support to producer organisations in countries in the global South. In the past he worked as a consultant for the Management and Development Foundation, and as a lecturer at both the Institute of Social Studies and Wageningen University. His PhD field work was undertaken in the Philippines and looked at water governance. He recently joined Twijnstra Gudde Management Consultants as a senior consultant on organisation development and change management.

Mr. Lucian Peppelenbos (PhD) joined KIT in 2005 as an advisor in sustainable chain development. Among his current projects are the development of an export chain in medicinal plants from India; the facilitation of a learning alliance in East Africa on pro-poor market development; the involvement of the Dutch private sector in poverty alleviation; and a global research project to identify best practices in connecting smallholder farmer to retailers and processors. In the past he worked in Chile as a freelance management consultant for agribusinesses, farmer cooperatives, and international agencies like Fair Trade and the Food and Agricultural Organization of the United Nations (FAO). He worked as a policy advisor in the Dutch Parliament, and at Agriterra, a Dutch non-profit organization for international cooperation between farmer organizations. He holds a PhD in Technology and Agrarian Development from Wageningen University. His expertise is in chain development, management optimization, and

business strategy development. He has professional skills in operational research, participatory approaches, multi-stakeholder processes, and training.

Ms. Rhiannon Pyburn (MSc) is a PhD candidate in the Communication and Innovation Studies Group at Wageningen University in the final stages of writing her doctoral dissertation, provisionally entitled: Reflexive Certification - Smallholder group certification as a development tool in the global South. Her field work was in Brazil, Thailand, Burkina Faso, Costa Rica and in Uganda. She is a consultant on issues related to social learning processes and social and environmental certification (e.g. organic agriculture and fair trade). Previous to her MSc and PhD work in the Netherlands, she worked in Indonesia, Côte D'Ivoire, across Canada and in the United States facilitating development and environmental education programs and cultural exchanges for youth. Her BSc from the University of Toronto (1995) was in International Development Studies - Resource Management.

Mr. Ted Schrader (MSc) is a development sociologist who, over the past 20 years, has been working in the fields of agricultural development and natural resource management in Africa. Currently he is liaison office for Agriterra and advises farmers' organisations in selected countries in East, Central and West Africa. In addition, he is involved in the further development of Agriterra's advisory services to producer organisations. His fields of interest include: organisational strengthening and institutional development, farmer-led agribusiness development, and the development of user-friendly methods and tools for farmers' organisations.

Mr. Giel Ton (MSc) works at the Agricultural Economics Research Institute (LEI) of Wageningen University and Research Centre as an agricultural economist. He is specialized in the analysis of institutional dynamics affecting smallholder livelihoods in developing countries and the role of farmers' organisations in value chain development. He has been working between 1989 and 1997 in Nicaragua developing economic service provisioning by smallholder organisations in Condega. From 1999 to 2004 he worked as policy advisor for the national farmers' federation CIOEC-Bolivia on issues related with the National Dialogue on the Bolivian Poverty Reduction Strategy and with regional and international trade negotiations.

Ms. Olga van der Valk (MSc) has an extensive track record working with producer groups on market integration and development. She worked as a marketing director and consultant for coffee cooperatives in Chiapas, Mexico for 13 years. After her return to the Netherlands in 2003, she conducted a market study on low cost water filters for the poor and supported the NGO platform - Netherlands Water Partnership - in developing a protocol for Appropriate Technology (now known as AT@work). Since 2004 she has been working as a researcher on international supply chains at the Agricultural Economics Research Institute (LEI) addressing issues related to group organization, quality standards and multi-stakeholder processes. She has an interest in the marketing integration of (small) farmers from an organizational learning perspective. In the two years with LEI she has worked with farmers and other stakeholders in the Czech and Slovak Republics, Ethiopia, Indonesia, Kenya, Thailand, Turkey, Uganda and Vietnam.

Producer organisations and market chains

315

Mr. Sietze Vellema (PhD) is senior scientist and program leader in the field of Chains, Innovation and Pro-poor Development at the Agricultural Economics Research Institute (LEI) at Wageningen University and Research Centre. He is involved in strategic policy research focusing on organization models, governance and management styles in cross-border food chains. His teaching and academic research activities are based at the Technology and  Agrarian Development group at Wageningen University.

Mr. Hugo Verkuijl (MSc) is an economist with expertise in institutional analysis; value chain analysis; the privatization of agricultural services; feasibility studies and impact assessment of privatization on the stakeholders; economic and policy research; and staff development. He has 15 years experience in international postings in Ethiopia and Mali and numerous short-term assignments in the Caribbean, Central America,  Eastern and Southern Africa and India. Currently, Mr Verkuijl is CEO of Mali Biocarburant: a pro-poor private enterprise that produces biodiesel from jatropha.

Mr. Bertus Wennink (MSc) joined the Royal Tropical Institute (KIT) in 1994, which entailed long term assignments and short term expert missions for different donors, NGOs and research institutions in West and Central Africa. Since 2003, he has been based in Amsterdam as a senior researcher (applied and action research), advisor and trainer for KIT. His speciality is multi-stakeholder approaches for institutional development, management of  demand-driven agricultural services, capacity strengthening of farmer organizations, community driven development, and natural resource management. He has additional expertise in the design and use of participatory approaches and tools and is an experienced trainer and workshop facilitator with strong intercultural communication skills.

Producer organisations and market chains

Ms. Gerda Zijm (MSc) worked from 1998 until 2004 for the Netherlands Development Organisation SNV. From 2001 she lead the programme for strengthening economic grassroots organisations in the Bolivian Valleys. Before that she worked as organisational advisor in several farmers' organisations in Nicaragua and Bolivia. Currently she works as policy advisor on rural planning for the Province of Utrecht in the Netherlands.

Producer organisations and market chains

Printed in the United States
by Baker & Taylor Publisher Services